Neutrinoless Double Beta Decay

Neutrinoless Double Beta Decay

Editors
V.K.B. Kota
U. Sarkar

Narosa Publishing House
New Delhi Chennai Mumbai Kolkata

Neutrinoless Double Beta Decay
164 pgs | 3 tbs | 31 figs

Editors
V.K.B. Kota
U. Sarkar
Theoretical Physics Division
Physical Research Laboratory
Ahmedabad, India

NAROSA PUBLISHING HOUSE PVT. LTD.

22 Delhi Medical Association Road, Daryaganj, New Delhi 110 002
35-36 Greams Road, Thousand Lights, Chennai 600 006
306 Shiv Centre, D.B.C. Sector 17, K.U. Bazar P.O., Navi Mumbai 400 703
2F-2G Shivam Chambers, 53 Syed Amir Ali Avenue, Kolkata 700 019

www.narosa.com

Printed from the camera-ready copy provided by the Editors.

ISBN 978-81-7319-876-2

Published by N.K. Mehra for Narosa Publishing House Pvt. Ltd.,
22 Delhi Medical Association Road, Daryaganj, New Delhi 110 002

Printed in India

FOREWORD
Probing the Two Greatest Mysteries of the Universe

D.P. Roy

Homi Bhabha Centre for Science Education, Tata Institute of Fundamantal Research,
Mumbai - 400088, India
Instituto de Fisica Corpuscular, CSIC-U de Valencia,
Correos E-46071 Valencia, Spain

One of the main achievements of modern physics is the unification of the microcosm with the macrocosm. It follows from the basic principles of uncertainty and mass energy equivalence that when we probe deep into the subatomic space we come across states of very high energy and mass. These are the particles that abounded in the early history of the universe, when its ambient temperature was as high as their mass. We have now a good understanding of these subatomic particles up to a mass range of 100 GeV in terms of the so called standard model. This has helped us to trace back the history of the universe to within a few picoseconds of its creation, when the ambient temperature was about a 100 GeV. Moreover we are now probing physics beyond the standard model mass scale of 100 GeV and looking into the very early history of the universe, when it was much less than a picosecond old. In the process we hope to understand the two greatest mysteries of the Universe, originating from that very early history – i.e. (1) the baryon asymmetry and (2) the dark (or invisible) matter.

Baryon Asymmetry of the Universe (BAU)

The visible universe is made of baryon (proton, neutron) and lepton (electron) with hardly any evidence of their antiparticles. This is a mystery since the standard model (SM) treats them symmetrically. As per the SM the basic constituents of matter are a dozen of fermions along with their antiparticles; i.e. the three pairs of leptons

$$\begin{pmatrix} \nu_e \\ e \end{pmatrix} \begin{pmatrix} \nu_\mu \\ \mu \end{pmatrix} \begin{pmatrix} \nu_\tau \\ \tau \end{pmatrix}$$

and three pairs of quarks, which constitute the baryons. They have three basic interactions – strong, electromagnetic and weak. These are all mediated by gauge bosons with couplings proportional to the corresponding gauge charges – i.e. colour charge, electric charge and weak isospin. Parity is broken maximally by the weak, but conserved by strong and electromagnetic interactions. Accordingly only the left-handed leptons and quarks occur as weak isodoublets, carrying isospin 1/2, as represented by the above pairs. On the other hand the right-handed charged leptons and quarks occur as isosinglets, carrying no isospin. It is exactly the opposite for the antiparticles, since all these gauge interactions preserve CP symmetry. Since the neutrino has no strong or electromagnetic interaction, there is no right-handed neutrino in the SM.

The mass term corresponds to a coupling of left- and right-handed fermions, which does not conserve isospin. Thus in the SM the matter fermions get their masses from spontaneous breaking of the weak gauge symmetry by the Higgs mechanism, which also gives mass to the weak gauge bosons. Therefore all these masses are limited by the symmetry breaking scale of about 100 GeV. But the neutrino has no mass in the SM since there is no right-handed neutrino. However, we know from the neutrino oscillation experiments now that the neutrinos have tiny but nonzero mass, much less than an eV, i.e. more than a billion times smaller than the other fermion masses.

A simple and elegant explanation of such tiny neutrino masses comes from the See-Saw model. It goes a step beyond the SM and assumes that the neutrino has a right-handed singlet component like the other fermions. Unlike the other fermions, however, the singlet neutrino is seen to have a unique property that it carries no gauge charge. Thus it is identical to its antiparticle, except for their opposite chiralities. This is called a Majorana fermion. It has a Majorana mass from the coupling of the left-handed neutrino with its right-handed antiparticle. Of course it breaks lepton number conservation, i.e. $\Delta L = 2$. But since the lepton number is not a gauge charge of the SM, this does not break any symmetry of the SM. So the Majorana mass M can be much larger than the SM symmetry breaking scale, i.e. $M >> 10^2$ GeV. Moreover, like the other fermions one also gets a normal (Dirac) mass m from the coupling of the right-handed singlet with the left-handed doublet neutrinos, which is limited by this symmetry breaking scale. Thus in the basis of left and right chirality one gets a 2x2 neutrino mass-matrix

$$\begin{pmatrix} 0 & m \\ m & M \end{pmatrix}$$

with $M >> m$. On diagonalization one gets the two mass eigenvalues as M and m^2/M. In contrast the other fermions have no Majorana mass, so that on diagonalization both the right and left handed fermions have the same Dirac mass m. So it is the more than a billion times larger Majorana mass of the right-handed singlet neutrino, that is responsible for pushing down the left-handed doublet neutrino mass more than a billion times lower than the other fermion masses. This is why it is called the See-Saw model. Thus the measurement of the ultralight neutrino masses in the neutrino oscillation experiments provides us indirect evidence for the ultraheavy Majorana masses, which can not be probed directly in any foreseeable future. Finally the lepton number violation, $\Delta L = 2$, associated with the ultraheavy Majorana mass can be responsible for generating a lepton asymmetry in the very early history of the universe, when its age was many billion times smaller than a pecosecond. Then the quantum anomaly associated with the SM could convert a part of this lepton asymmetry into a baryon asymmetry at a slightly later stage, when its age was a few picoseconds. This is in fact the leading model for explaining the BAU, which is evidently a very important matter as it underlies our very existence in the universe. But we have still a long way to go in order to prove (or disprove) this model.

Figure 1 illustrates what we know about the masses and mixings of the three light neutrinos from atmospheric and solar neutrino oscillation data along with those from accelerator and reactor neutrino experiments (E. K. Akhmedov, hep-ph/0610064).

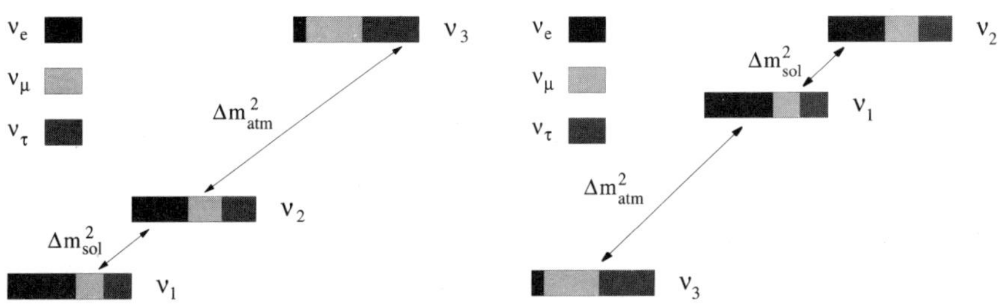

Figure 1. A schematic diagram of nutrino masses and mixings

There are two mass-square gaps Δm^2 between the three neutrino masses. Thanks to the solar matter effect, we know both the magnitude and the sign of the smaller gap $m_{sol} \simeq 8 \times 10^{-5}\ eV^2$, while we know only the magnitude of the larger gap $m_{atm} \simeq 2.5 \times 10^{-3}\ eV^2$. We also know two of the three mixing angles, which are both large; but we have only a limit on the third one representing ν_e content of the 3rd mass eigenstate, i.e. $|U_{e3}^2| < 0.05$. The sign of the m_{atm}^2 leads to a two-fold ambiguity called normal and inverted hierarchies, as illustrated above. Secondly we do not know the absolute mass scale – the lowest mass could be larger than the gaps implying relatively large and degenerate neutrino masses. Third and most important, we do not know for sure the Majorana nature of these masses, since the neutrino oscillation involves no L violation. So we can not rule out the possibility that the neutrino has no Majorana mass, but its Dirac mass is much smaller than those of other fermions for some as yet unknown reason.

The neutrinoless double beta deacay (NDBD) experiment holds the key to all these three issues. Many experiments have observed double beta decay between neighbouring even-even nuclei,

$$^{A}_{Z}X \to {}^{A}_{Z+2}X + 2e^- + 2\bar{\nu}_e$$

If neutrinos have Majorana mass then the associated lepton number violating mechanism can lead to pair-annihilation of the two antineutrinos, leading to the NDBD

$$^{A}_{Z}X \to {}^{A}_{Z+2}X + 2e^-,$$

which will show up as a peak in the total kinetic energy of the electron pair. The rate of the NDBD is proportional to the square of the Majorana mass

$$|< m_{ee} >| = |\sum_i m_i U_{ei}^2|,$$

where the constant of proportionality comes from the nuclear matrix elements. They can be estimated from processes like (1), although there is still considerable uncertainty in their estimates. The NDBD experiment of the Moscow-Hedelberg collaboration has already limited the above Majorana mass to $< m_{ee} >< 0.35$ eV. Assuming Majorana nature of neutrino mass this gives arguably the best limit on the mass of each neutrino in the degenerate mass scenario. In the inverted hierarchy scenario the two ν_e rich mass eigenstates of the above figure have atmospheric scale mass of 0.05 eV. The next generation of NDBD experiments propose to extend the probe of neutrino Majorana mass down to this level, i.e. they can cover the inverted hierarchy along with the degenerate mass scenarios. If they find no NDBD signal then the final step will be to extend the probe down to the solar neutrino mass scale of about 0.01 eV, corresponding to the mass of the ν_e rich states in the normal hierarchy scenario. At this level the NDBD experiment will be able to either confirm or disprove the Majorana nature of neutrino mass. This last step may be a long way off; but there is no other way for this lofty goal.

Dark Matter of the Universe (DM)

We know for a long time now that the ordinary baryonic matter constitutes only a small fraction of the matter content of the universe, while most of it comes from an invisible matter of unknown origin, which is called the dark matter (DM) for historical reason. The oldest and the strongest evidence for DM comes from the rotational velocity curves of stars or hydrogen clouds, lying outside the visible disc of galaxies, as shown in the following figure.. Balancing the centrifugal acceleration with the gravitational attraction of the galaxy gives

$$\frac{v^2}{R} = \frac{G_N M(R)}{R^2} \Longrightarrow v^2 = \frac{G_N M(R)}{R}$$

If there were no galactic matter outside the visible disk then the rotation velocity curve would have gone down like $1/\sqrt{R}$. Instead it continues to rise way beyond the visible disk, suggesting that there is a lot of invisible matter in and around the galaxy. Similar results have been obtained from about a thousand galaxies, including our own, as well as cluster of galaxies.

A natural candidate for this invisible matter comes from supersymmetry, which is invoked to stabilise the symmetry breaking scale of the SM against divergent quantum corrections. It is the lightest suppersymmetric particle (LSP), which is a colourless and chargeless object having only weak interaction like the neutrino, but with a mass of 10^{2-3} GeV. It is also called weakly interacting massive particle (WIMP). This particle can be looked for at the forthcoming large hadron collider at CERN, Geneva, at least over a part of the supersymmetric parameter space. But whether such particles are seen at CERN or not, their connection with the invisible or dark matter of the universe can

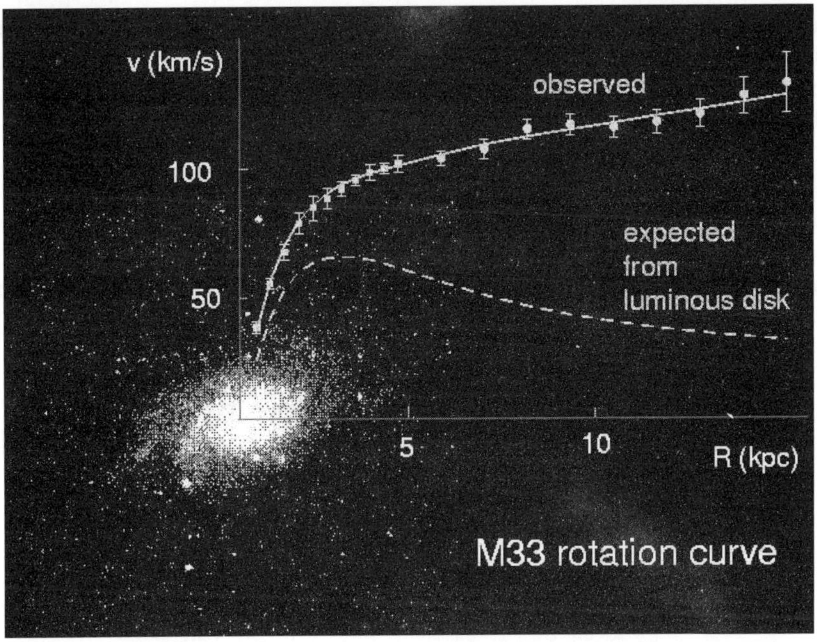

Figure 2. Rotation velocity curve of the nearby dwarf spiral galaxy M33, superimposed on its optical image (E. Corbelli and P.Salucci, astro-ph/9909252)

only be established through the direct dark matter detection (DDMD) experiments. It is based on measuring the nuclear recoil energy in the elastic scattering of the DM on a heavy nucleus. Since the recoil energy comes from the kinetic energy of the DM, which is only 10^{1-2} keV, one needs to have a very low energy threshold.

Concluding Remarks

As we have seen above the NDBD and the DDMD experiments probe the origins of the two greatest mysteries of nature, i.e. the baryon asymmetry and the dark matter of the universe. They are both challenging experiments. The reaction rates are very low, which calls for large target mass. Besides one has to contend with high cosmic ray and radioactive backgrounds, which call for deep underground location and selection of extremely radiopure materials and environment. Not withstanding these challenges, there are about a dozen of DDMD experiments around the world and about twice as many NDBD experiments. This shows the degree of global importance and priority enjoyed by these two experiments. It may be noted here that these challenges being common to both NDBD and DDMD experiments, very often they face it together by sharing the same underground laboratory and the same radiopure environment. Moreover some of the NDBD experiments, which are based on bolometric technique, have been able to achieve sufficiently low energy threshold and carry on DDMD simultaneously. I have had the opportunity to see two such experiments. One was the abovementioned

Moscow-Heidelberg experiment in the Gran-Sasso road tunnel in Italy, which has simultaneously looked for both NDBD and DDMD. The second was the dark matter research centre in a Hydroelectric power tunnel in South Korea, which is engaged in both these searches.

Finally I compliment the organisers of this dedicated workshop on NDBD and DDMD to have brought together theorists and experimentalists from India and abroad to take stock of the present status and future prospects of these two very important experiments. And I conclude with the hope that this will help to advance the nascent effort of the Indian physics community to initiate these experimental activities in this country.

FOREWORD

Is Neutrino a Majorana Particle?

G. Rajasekaran[1]

Institute of Mathematical Sciences, Madras 600113, India

Dirac introduced the concept of antiparticles while trying to solve the negative energy problem in the famous relativistic equation for the electron that he had discovered in 1928. Now we know that the existence of antiparticles is one of the most important consequences of combining quantum mechanics with special relativity. For every particle there exists an antiparticle. However some particles could be self-conjugate, in the sense that particle and antiparticle could be the same. Of course such particles have to be electrically neutral. Among the elementary particles of the Standard Model, photon and the Z boson are self-conjugate. Both these are bosons. The possibility of a self-conjugate fermion was first pointed out by Majorana in 1937 and hence they are called Majorana fermions while the other fermions (with distinct particles and antiparticles) are called Dirac Fermions. Among the fermions of the Standard Model, only neutrinos are electrically neutral and hence qualify to be Majorana particles. But it is still an open question whether neutrinos are Majorana particles or Dirac particles, in other words, whether neutrino is a self-conjugate particle or not.

This question can be shown to be a pure semantic one if neutrinos are massless. In that case one can prove, by renaming the left-handed neutrino and the right-handed antineutrino suitably, that there is no physical distinction between Dirac and Majorana neutrinos. This is somctimes called "the confusion theorem". But after the discovery of neutrino mass in the last decade, we know that such a distinction exists and physicists must determine the category to which neutrino belongs.

There are important reasons why theoreticians prefer Majorana neutrinos. If neutrinos are Majorana particles, there exists an elegant mechanism called sea-saw to explain why the neutrino masses, although not zero, are so tiny. Further, if neutrinos are Majorana particles, lepton number L is not conserved and this opens the door to

[1] **E-mail:graj@imsc.res.in**

generate an excess of leptons over antileptons in the early universe which can subsequently generate an excess of baryons over antibaryons, thus explaining how after annihilation of most of the particles with antiparticles, a finite but small residue of particles was left, to make up the present Universe. Hence the fundamental importance of the question whether neutrinos are Majorana particles is clear.

Inspite of the great attractiveness of the idea that neutrinos could be Majorana particles, it is only a theoretical idea. It has to be either confirmed or refuted by experiment. At the present stage, the only experiment that can answer this question is the neutrinoless double beta decay (NDBD). In neutrinoless double beta decay,the two Majorana neutrinos that are virtually produced can annihilate each other leaving only two electrons in the final state, thus violating lepton number by two units. So establishment of the existence of this decay would be proof of the Majorana nature of the neutrino and violation of lepton number.

Usual double beta decay (in which two neutrinos along with two electrons are emitted) itself is a very rare process because it is doubly weak as compared to the standard beta decay and further the phase space is suppressed by the larger number of particles in the final state. Inspite of its rarity, this decay has been experimentally detected and now well studied. However as we saw above, the signature for the Majorana character of the neutrino is the absence of the two neutrinos and this process without neutrinos in the final state has not yet been detected. Although the neutrinoless double beta decay has reduced number of particles in the final state and hence more phase space, it is a much rarer process than the two neutrino decay because the decay amplitude is proportional to the tiny neutrino mass (consistent with the confusion theorem).

Actually the decay amplitude for neutrinoless double beta decay is proportional to a neutrino mass factor m_{ee} which is a linear combination of the masses of the three neutrinos. This linear combination also involves the neutrino mixing angles and the very important CP violating phases. Of course this has to be multiplied by the relevant nuclear matrix element to get the decay amplitude. Thus if neutrinoless double beta decay is detected and the rate is measured and if the nuclear matrix element is known, one can then extract important information on the neutrino parameters. Especially to be noted is the fact that m_{ee} contains the overall scale of the neutrino masses and this mass-scale is not obtainable from neutrino oscillation experiments. However the success of such an extraction of neutrino parameters from NDBD depends very crucially on our knowing the nuclear matrix elements and hence nuclear theory will play an essential role in NDBD activity.

Can the NDBD neutrino mass factor m_{ee} be zero? Unfortunately the answer is yes. The particular linear combination of neutrino masses that enters here may be zero (or very small) for some reason and then the rate for NDBD may be vanishingly small. In this unfortunate case, NDBD experiments will not throw any light on whether neutrinos are Majorana particles. One may have to go to processes such as a μ^- colliding with a nucleus leading to a μ^+ in the final state. The amplitude for this reaction is proportional to a different linear combination of neutrino masses which hopefully is not zero. But this lepton number violating reaction will be even harder to study than NDBD.

Is it possible to enhance the rates for the lepton number violating processes by some mechanism, for instance by shining an intense laser? Such ideas are being considered and one of them might work. But for the present let us return to NDBD.

Inspite of the many experiments that have been mounted for searching for neutrinoless double beta decay, none has borne fruit sofar. However,in 2004 Klapdor and his group reported observing the decay in ^{76}Ge. This created quite an excitement because of the importance of such a discovery. However this result was soon controverted by many critical physicists. I was reading Galileo's biography when this news came. My mind went back by 400 years to the time when Galileo was facing his sceptical opponents who refused to believe that Galileo really saw the things that he claimed to have seen through his spy glass. Of course times have changed and the situation is quite different. In any case there is no religious dogma concerning the nature of the neutrino! It is very important to settle the issue by independant experiments.

The National Workshop on "Neutrinos in Nuclear,Particle and Astrophysics (NUPA04)" held at IIT,Kharagpur in Feb 2004 was the first such workshop covering the impact of the recent discoveries of Neutrino Physics in all the three fields of Nuclear Physics, High Energy Physics and Astrophysics. In that Workshop I stressed the importance of the neutrinoless double beta decay experiment and suggested that the Double Beta Decay Theorists Prof P K Raina and Prof P K Rath must organize and coordinate the activity that could lead to mounting a Neutrinoless Double Beta Decay Experiment in the INO cavern. Their magnificent enthusiasm led to two focussed Workshops DBD05 at IIT,Kharagpur in March 2005 and NDBD05 at Lucknow in November 2005. More importantly, they have succeeded in bringing two excellent experimenters Vivek Datar and Pillay and their groups into the NDBD project and this augurs well for the ultimate success of the project.

Utpal Sarkar and Kota must be congratulated for bringing the experimenters and theorists together for the NDBD07 at Ahmedabad in February 2007 and for bringing out the proceedings. The articles will serve as status reports of the NDBD activities sofar and help us to plan for the future.

Let me close this foreword with the following remarks relevant to the roadmap ahead:

1. The fundamental importance of neutrinoless double beta decay must be stressed again and again. Repetition of this manthra is not a waste of time since the most important unknown in all of neutrino physics is the answer to the question whether neutrino is Majorana or Dirac. The answer will have a bearing on High Energy Physics as well as Cosmology. At the present, NDBD experiment is the only way to answer it.

2. Hence at the India-based Neutrino Observatory (INO), NDBD must be pursued with full vigour. Although the major focus of the INO as of now is to construct the magnetised iron calorimeter (ICAL) to be used for atmospheric neutrinos (Phase I) and neutrinos from muon storage rings and beta beams (Phase II), parallel processing of NDBD must go on. NDBD activity must soon gain sufficient strength so that the NDBD detector will become as (if not more) important as ICAL to INO. But this requires considerable spadework and R and D.

3. It is absolutely essential to scout for good experimenters and augment the NDBD group. We must scan the whole spectrum of likely candidates in research institutions and universities inside and outside the country and attract them to NDBD. Also

we must not restrict ourselves to HEP and NP experimenters only. We must interact with atomic physicists, condensed matter physicists, material scientists, chemists, engineers... All of these can make useful contributions to mounting a viable NDBD experiment in the country. Students must form an important component of the human resource that we seek for NDBD.The task is a gigantic one, but it can be done and it must be done.

4. Finally we must mention that experiments on the search for dark matter have reached very high sensitivities and are becoming capable of detecting it, if it really exists in the form and abundance generally expected. Hence we must include the possibility of the Indian NDBD project leading to an Indian Dark Matter project in the future.

I thank VKB Kota and Utpal Sarkar for asking me to write the foreword.

PREFACE

One of the most important questions in neutrino physics is, whether neutrinos are Majorana particles and hence there is lepton number violation in nature. This may have consequences in astrophysics and cosmology. Search for the neutrinoless double beta decay (NDBD) could answer this question and provide the Majorana mass of the neutrinos, which is the most important unknown parameter in neutrino physics at present. In this effort the main problem is to detect this extremely rare process. In addition, proper interpretation of the results require calculation of complicated nuclear structure matrix elements from nuclear models. Considering the importance, many new experiments are planned all over the world.

In India there is a sincere effort to perform this experiment in the same underground laboratory as India-based Neutrino Observatory. Feasibility study is going on at present in places such as TIFR, BARC, IMSc, SINP, VECC, IIT-Kharagpur and Lucknow University, in which PRL is taking an important role in theoretical aspects. The purpose of the present meeting is reviewing the status of our progress and plan future activities. There were lectures on different aspects of NDBD, present status of all NDBD experiments and developments with the Indian proposal for NDBD. A detailed preview of the present meeting and also summaries of the DBD meetings held in 2005 at IIT-Kharagpur and Lucknow University are given in the introductory article in the proceedings. Since this is the first proceedings that discuss the possibility of the neutrinoless double beta decay experiment and also mentions about the possibility of the dark matter search experiment in India, we hope that this proceedings become useful to many readers.

We thank Director of PRL, Prof. J.N. Goswami, for inaugurating the workshop, supporting this activity and providing financial support for holding the workshop and bringing out the proceedings. We also thank the PRL administration for all their support in organizing the workshop. We are specially indebted to Profs. D.P. Roy and G. Rajasekaran for readily agreeing to write forewords to the proceedings and for their inspiration to this activity. Last but not the least, we thank all the participants who could come for this meeting in a very short notice and particularly the authors for submitting their articles in time. We hope that this proceedings will be helpful for those who would like to know the present status of the subject and a good reference for future research in double beta decay and related topics.

V.K.B. Kota and U. Sarkar
(Editors)

Ahmedabad

CONTENTS

NEUTRINOLESS DOUBLE BETA DECAY
Editors: V.K.B. Kota and U. Sarkar

INTRODUCTION

P. K. Raina[1], P. K. Rath [2], R. G. Pillay[3], V. Nanal[3], V. M. Datar[4], A. Srivastava[4], V.K.B. Kota[5] and U. Sarkar[5]

[1]Indian Institute of Technology, Kharagpur - 721302, India
[2] University of Lucknow, Lucknow - 226007, India
[3] Tata Institute of Fundamental Research, Bombay-400005, India
[4] Bhabha Atomic Research Center, Bombay 400085, India
[5] Physical Research Laboratory, Ahmedabad-380009, India

Abstract

The study of neutrinoless double beta decay has become indispensable due to the confirmation of flavor oscillation of neutrinos to pin down the mass and nature of the later. We present briefly the scope of this topic that is further elaborated in subsequent articles of this volume. The theoretical as well as experimental efforts taken in India over the past two years to set up an experiment to study neutrinoless double beta decay are also reported and this includes a brief report of the two previous meetings held at Kharagpur and Lucknow.

1. Introduction

The study of "Neutrino Physics" in general and "Neutrinoless Double Beta decay" in particular is one of the present day thrust areas in experimental as well as theoretical physics. In 1930, Wolfgang Pauli postulated the existence of electron neutrino ν_e – a chargeless, massless and spin 1/2 particle – to explain the continuous nature of beta spectrum, and Fermi gave the theory of beta decay, almost immediately, assuming the electron neutrino to be a Dirac particle. In 1955, Cowan and Reines experimentally detected neutrinos at the Savannah river reactor site. In 1956, T. D. Lee and C. N. Yang suggested that parity is not conserved in weak interaction, and Madam C. S. Wu established experimentally that the weak interaction in fact violates the conservation of parity maximally. With this discovery, it became apparent that the ν_e is a Weyl particle, i.e. chargeless, massless and left-handed spin 1/2 fermion. In the mean time, two other flavors of neutrinos, namely muon neutrino ν_μ and tau neutrino ν_τ have been discovered. In the paradigm of Standard Model of electroweak unification (SM), which asserts that the material universe consists of four fundamental forces and forty-five chiral fermions, all the three neutrinos are assumed to be Weyl particles.

The neutrino is the most enigmatic member of the particle zoo. Neutrinos are extremely difficult to detect because they hardly interact with matter – only one in a million of the neutrinos traversing Earth collide with an atomic nucleus. In 1967, Raymond Davis, using a huge tank of cleaning fluid (HCl_4) placed deep underground to shield it from cosmic rays, started looking for neutrinos produced by the nuclear reactions in the Sun's core. However, he could only detect about one third of what was expected. Other underground Gallium detectors, namely SAGE, GALLLEX and GNO found similar neutrino shortages and the magnitude of deficit varied roughly between one-half to one third. Bruno Pontecarvo in 1957 and independently, Maki, Nakagawa and Sakata in 1962, proposed that neutrinos might change from one flavor into another through flavor oscillation [1]. Gradually, physicists started to suspect that oscillation could explain the deficit of neutrinos detected from the Sun. The detectors could capture only one type of neutrino, and neutrinos during their journey from the Sun changed flavor and escaped detection. Similar deficiencies were also observed in the experiments detecting atmospheric neutrinos.

Finally, the evidence for oscillation of atmospheric and solar neutrinos came from Kamiokande and later, this evidence was confirmed by Super-Kamiokande experiment with better statistics [2]. The detection of separate neutrino flux from electron neutrinos ν_e through charged current reactions, and the flux of all active neutrinos through the neutral current reactions at SNO, suggested that the solar neutrino oscillation is inescapable [3]. An overview of the SNOLAB has been presented by **R. U. Haq** in this volume. The K2K long base line probes the ν_μ disappearance and provides the first confirmation of atmospheric neutrino oscillation in the laboratory [4]. It has been further observed at the KamLAND reactor experiment that $\overline{\nu}_e$ oscillates as well [5] and the oscillation parameters are in agreement with those of solar neutrino experiments. This agreement based on CPT invariance shows that the matter effect on oscillation is well understood. The CHOOZ and Palo Verde reactor experiment data imply that the angle θ_{13} is extremely small [6, 7]. The study of SN1987A explosion, WMAP, SDSS and 2dFGRS has further supplemented information on mass and nature of neutrinos.

The confirmation of neutrino oscillation in atmospheric, solar, reactor and accelerator experiments established the fact that neutrinos have mass. However, neutrinos are assumed to be massless in the SM. Hence, some new physics beyond the SM has been already hinted. The massive neutrinos may be of Dirac or Majorana character. The mass and nature of neutrinos play an important role in theories beyond the SM, namely GUTs, supersymmetry, supergravity and string theory. Further, the neutrinos play an important role in stellar evolution, supernova explosion, galaxy formation, thermal history of the universe, big-bang nucleosynthesis, dark matter and dark energy. The data from neutrino oscillations provide us with mass squared differences, mixing angle and possible mass spectrum of neutrinos. There are a number of possibilities to study lepton number violation and Majorana character of neutrinos. A detailed discussion on processes implying lepton number violation and Majorana character of neutrinos can be found in the article by **S. D. Rindani**. However, it is generally agreed that the mass and nature of neutrinos (Dirac or Majorana) can be extracted only through the study of single beta decay and neutrinoless double beta decay (NDBD) together.

It is generally agreed that our material universe consists out of 5% of baryonic matter, 25% of dark matter and 70% of exotic dark energy. Further, it is suggested through astronomical observations that most of the mass in the Galactic halo is dark and the dark matter consists of some new, as yet undiscovered weakly interacting massive particle (WIMP). Out of many possibilities, the best theoretically motivated WIMP candidates are the light stable sparticles of supersymmetric theories and this speculative supersymmetric dark matter based on simple physical principles is well motivated. Experiments devoted to study NDBD can also be employed to detect WIMPS in the galactic halo. The most promising techniques are direct detection in the low-background laboratory detectors and indirect detection through observation of energetic neutrinos from annihilation of WIMPS that have accumulated in the Sun and/or the Earth. There is already some evidence regarding the detection of WIMPS and further details on the experimental detection of WIMPS can be found in the article by **R. Bernabei**. Thus, the study of NDBD has important consequences in nuclear physics, particle physics, astrophysics as well as cosmology. Further, it is going to open up new research frontiers in theoretical physics as well as add new technological know-how to the experimental physics.

2. Double Beta Decay and its Implications

The nuclear double beta ($\beta\beta$) decay is a rare and spontaneous second order process involving weak interaction and takes place between two even Z-even N isobars differing in charge by two units. Historically, the first clue to nuclear $\beta\beta$ decay was that some even Z-even N nuclei, which are otherwise candidates for β decay and should have become extinct through radioactive disintegrations, are still found in the earth's crust billions of years after their formation. The observed unexpected stability of these even Z-even N nuclei was explained by Werner Heisenberg by postulating the existence of a pairing force between like nucleons, due to which these even-even nuclei are more bound than the neighboring odd-A nuclei and stable against β decay. However, this argument for the stability of these even Z-even N nuclei is not complete in itself because, if a particular even-even nucleus is lighter than the nearest odd-odd nucleus but heavier than the next nearest even-even nucleus then $\beta\beta$ decay should indeed occur. The two neutrino double beta ($2\nu\ \beta^-\beta^-$) decay process given by

$$ {}^{A}_{Z}X \rightarrow {}^{A}_{Z+2}X + 2e^- + 2\overline{\nu}_e \qquad (1) $$

was studied by Maria Goepert-Mayer [8] at the suggestion of Eugene P. Wigner to explain the observed stability of some even-even nuclei. The half-life of $2\nu\ \beta^-\beta^-$

decay was calculated by Mayer and it was greater than 10^{17} yr, which is exceedingly slow even on the geological time scale.

In 1939, W. Furry conjectured another decay mode, namely electron emitting neutrinoless double beta ($0\nu\ \beta^-\beta^-$) decay [9] assuming the ν_e to be a Majorana (particle≡antiparticle) particle [10, 11].

$$^{A}_{Z}X \rightarrow {}^{A}_{Z+2}X + 2e^- \tag{2}$$

In $0\nu\ \beta^-\beta^-$ decay, there is no corresponding requirement for neutrinos, although the conservation of energy and momentum required the emission of a neutrino in single β decay. The energy and momentum could be conserved in this decay mode releasing two electrons only. Thus, the $2\nu\ \beta\beta$ and $0\nu\ \beta\beta$ decay modes are energetically competing processes. The mechanism of $0\nu\ \beta^-\beta^-$ decay is based on the emission of an electron antineutrino $\overline{\nu}_e$ on the first decay vertex ($n \rightarrow p + e^- + \bar{\nu}_e$) and its absorption in the second vertex. The absorption part is an inverse beta decay ($n + \nu_e \rightarrow p + e^-$) where an electron neutrino ν_e instead of the antineutrino $\bar{\nu}_e$ is required. This lepton number violating process is possible only if the neutrino is a Majorana particle. Furry estimated the half-life of a nucleus in the $0\nu\ \beta^-\beta^-$ decay mode for the first time.

The nuclear $\beta\beta$ decay is also possible through the emission of positrons. There are three possible modes, namely double-positron ($\beta^+\beta^+$) emission, electron-positron conversion (β^+EC) and double-electron capture ($ECEC$), which are given as follows.

(i) Two neutrino positron decay modes

$$\begin{aligned} {}^{A}_{Z}X &\rightarrow {}^{A}_{Z-2}X + 2e^+ + 2\nu_e \\ e^- + {}^{A}_{Z}X &\rightarrow {}^{A}_{Z-2}X + e^+ + 2\nu_e \\ 2e^- + {}^{A}_{Z}X &\rightarrow {}^{A}_{Z-2}X + 2\nu_e \end{aligned} \tag{3}$$

(ii) Neutrinoless positron decay modes

$$\begin{aligned} {}^{A}_{Z}X &\rightarrow {}^{A}_{Z-2}X + 2e^+ \\ e^- + {}^{A}_{Z}X &\rightarrow {}^{A}_{Z-2}X + e^+ \\ 2e^- + {}^{A}_{Z}X &\rightarrow {}^{A}_{Z-2}X^* \end{aligned} \tag{4}$$

The later three processes are energetically competing and we refer to them as positron double beta decay (e^+DBD) modes (refer to articles by **P. K. Raina** and **M. Saran** for more details). There are 35 possible candidates undergoing $\beta^-\beta^-$ decay and about equal number for e^+DBD.

The $2\nu\ \beta\beta$ decay modes conserve the lepton number exactly and hence it is allowed in the SM. The half-life of $2\nu\ \beta^-\beta^-$ decay is a product of accurately calculable phase space factor and model dependent nuclear transition matrix element (NTME). It has already been observed for the $0^+ \rightarrow 0^+$ transition in 11 nuclei, namely ^{48}Ca, ^{76}Ge, ^{82}Se, ^{96}Zr, ^{100}Mo, ^{116}Cd, ^{128,130}Te, ^{136}Xe, ^{150}Nd and ^{238}U, out of 35 possible candidates [12]. Hence, the values of NTMEs can be extracted directly. Consequently, the validity of different models employed for nuclear structure calculations can be tested by calculating NTMEs (more details can be found in the articles by **P. K. Rath, V. K.**

B. Kota, and **R. Sahu**; nuclear models are also essential for $0\nu\ \beta\beta$ decay as discussed in these articles).

On the other hand, the $0\nu\ \beta\beta$ decay, which is far more interesting and important as it violates the lepton number conservation due to the virtual exchange of Majorana neutrino, is not allowed in the SM. However, the $0\nu\ \beta\beta$ decay is possible in many gauge field theoretical models beyond the SM, namely GUTs - left-right symmetric SO(10) and E(6) models, R_p-conserving as well as violating SUSY models and higher dimension string theories. It is also possible in the scenarios of lepto-quark exchange, compositeness and sterile neutrinos. In many standard extensions of the SM, the $0\nu\ \beta\beta$ decay is possible through the emission of single ($0\nu\ \beta\beta\phi$) [13] or double Majorons ($0\nu\ \beta\beta\phi\phi$) [14] and are explicitly written as

$$
\begin{aligned}
{}^{A}_{Z}X &\rightarrow {}^{A}_{Z+2}Y + 2e^{-} + \phi && (5)\\
{}^{A}_{Z}X &\rightarrow {}^{A}_{Z+2}Y + 2e^{-} + 2\phi && (6)
\end{aligned}
$$

These $0\nu\ \beta\beta\phi$ and $0\nu\ \beta\beta\phi\phi$ modes are possible in nine Majoron models [15]. The existence of Majorons can play a crucial role in many areas, namely physics beyond the SM, evolution of stellar objects, supernovae astrophysics, history of the early universe and many other cosmological issues. Hence, the $0\nu\ \beta\beta$ decay is a convenient tool to test the physics beyond the SM in general and the best alternative to distinguish whether the neutrino is of Dirac or Majorana type in particular. Further details on the importance and implications of $0\nu\ \beta\beta$ decay can be found in the article by **U. Sarkar**.

Thus, the nuclear $\beta\beta$ decay is characterized by four experimentally distinguishable processes, namely $2\nu\ \beta\beta$ decay, $0\nu\ \beta\beta$ decay, $0\nu\ \beta\beta\phi$ decay and $0\nu\ \beta\beta\phi\phi$ decay. The latter three modes of $0\nu\ \beta\beta$ decay have not been observed so far. The best limit for the $0\nu\ \beta^-\beta^-$ decay half-life is $T^{0\nu}_{1/2} > 1.9 \times 10^{25}$ yr and it is derived for ^{76}Ge [16]. However, Klapdor and his group have reported recently that $0\nu\ \beta^-\beta^-$ decay has been observed in ^{76}Ge and $T^{0\nu}_{1/2} = 1.19 \times 10^{25}$ yr [17]. This result is controversial but it is expected that the issue will be settled soon. The present aim of all experimental activities is to observe the $0\nu\ \beta^-\beta^-$ decay. One can extract stringent limits on effective neutrino mass $\langle m_\nu \rangle < 0.35\ eV$, effective right handed weak current parameters $\langle \lambda \rangle < 6.21 \times 10^{-7}$, $\langle \eta \rangle < 3.44 \times 10^{-9}$ and the effective Majoron coupling constant $\langle g_M \rangle < 0.82 \times 10^{-4}$ from the observed limit on $T^{0\nu}_{1/2}$ in ^{76}Ge [16]. Stringent limits on R_p-conserving as well as violating SUSY parameter, lepto-quark coupling, coupling constant for compositeness model, sterile neutrinos and amount of violation of weak equivalence principle can also be obtained. However, the information about the extracted gauge theoretical parameters is limited by the uncertainty in the model dependent NTMEs.

To summarize, the significance of studying $0\nu\ \beta\beta$ decay lies in the fact that it directly addresses the following questions related to physics beyond the SM, namely

(i) origin of neutrino mass (Dirac vs. Majorana)

(ii) absolute scale on neutrino mass

(iii) type of hierarchy

(iv) CP violation in the lepton sector.

3. National Workshops on Neutrinoless Double Beta Decay

With the above mentioned perspectives in mind, a three day National Workshop on "Neutrinos in Nuclear, Particle and Astrophysics (NUPA04)" was held at IIT, Kharagpur from February 26-28, 2004. Various activities in the field of "Neutrino Physics" related to nuclear, particle and astrophysics were discussed. During discussions, it was realized that in spite of the fact that there were already about 30 "presently running" and "proposed" experiments all over the world to study the neutrinoless double beta decay (NDBD), there was no such experimental activity in India. However, there were proposals to study the neutrino oscillation at INO. G. Rajasekaran, D. P. Roy and Utpal Sarkar suggested that there should be efforts to consider the setup of experiments on NDBD in India utilizing the facilities to be available at INO. There was an intense discussion with the late Prof. C. V. K. Baba regarding the possibility of setting up an NDBD experiment in India. Finally, it was agreed that work should start on possible experiments in NDBD in India. It was also suggested that IIT, Kharagpur will be an appropriate institute to work as a nodal center to carry forward such activities. Subsequently, a meeting was held with the Director of the institute and the response was quite positive.

3.1. DBD05

In pursuance of the above mentioned proposal, a one day workshop on "Double Beta Decay (DBD05)" was held at IIT, Kharagpur on March 13, 2005. In the key note address, G. Rajasekaran elaborately discussed the nature of neutrinos, namely Dirac and Majorana character, origin of light mass left-handed neutrinos through "seesaw mechanism" and neutrino mixing. Further, he discussed the flavor oscillation of neutrinos in atmospheric, solar, accelerator and reactor experiments and the relevance of $\beta\beta$ decay experiments in the context of confirming Majorana nature of neutrinos. Subsequently, theoretical and experimental aspects of $\beta\beta$ decay were discussed by P. K. Rath and V. M. Datar respectively. Two specific experiments namely NEMO and CUORE as prototype of tracking and calorimetric experiments were presented by Aradhana Srivastava and Vandana Nanal respectively. The positron double beta decay experiments in general and Cobra type in particular was proposed by Ajay Singh. N. K. Mondal gave a beautiful overview of underground experiments in general and INO in particular. He also suggested that INO will be quite suitable to house the $\beta\beta$ decay experimental setup. To this end, V. M. Datar, who is also a member of INO, was requested to look into the planning and other related issues for the experimental setup of NDBD in future. Finally, five working groups, namely Sn-124 group, Nd-150 group, Positron group, Nuclear theory and Particle physics theory groups under the coordinatorship of R. G. Pillay, V. M. Datar, P. K. Raina, P. K. Rath and Utpal Sarkar respectively were formed through the collaboration of scientists from different national institutes and universities, namely TIFR, BARC, SINP, VECC, IMSc, PRL, NSC, IIT (Kharagpur), IOP and University of Lucknow under the overall coordrinatorship of P. K. Raina. Establishing of a small double beta decay group as a part of INO was discussed in the meeting and incorporated in the INO report [18]. Details of INO programme and its present status are given in the article by **V.M. Datar**.

3.2. NDBD05

A two day workshop on "Neutrinoless Double Beta Decay (NDBD05)" was held at University of Lucknow during November 4-5, 2005. The objective of the proposed workshop was to discuss the experimental as well as theoretical feasibility of definite experimental setups on NDBD. Further, it was expected that the deliberations of the workshop will bring to light novel techniques to be implemented in the experimental studies of NDBD which in turn is likely to prove to be a unique landmark in the international scenario of experimental physics.

The meeting started with an elaborate and beautiful presentation by D. P. Roy on mass and nature of neutrinos and implication on NDBD. Pomita Ghoshal presented results of atmospheric neutrino oscillation. The status of INO was presented by N. K. Mondal at length. The perspectives of NDBD was discussed by P. K. Raina in detail. The nuclear structure aspect and nuclear effects in neutrino induced processes were discussed by C. R. Praharaj and S. K. Singh respectively. Ramesh Chandra and Samata Singh discussed the study of $\beta\beta$ decay of ^{150}Nd and e^+DBD in the PHFB model. The discussion of experimental aspects of NDBD started with discussion of background issues by A. K. Singh. Vandana Nanal discussed the feasibility of ^{124}Sn bolometer for the study of 0ν $\beta\beta$ decay. Here below, we briefly discuss the main points mentioned by her.

In 0ν $\beta\beta$ decay, the decay rate is proportional Q^5. Hence, the $\beta\beta$ decay emitter should be of high Q value. This has an additional advantage as the interference from natural radioactive background is less at higher energies. The expected events are rare due to long half-life. Therefore, both the source and detector have to be large in size. In active experiments, in which the source and detector are the same, the experimental setup is simplified considerably. Another crucial requirement in the design of the detector is high energy resolution for a precision measurement of the sum energy of two electrons emitted in 0ν $\beta\beta$ decay. For this purpose, a bolometric detector is ideally suited. Out of the two candidates, namely ^{124}Sn and ^{150}Nd, only Sn can be made into a bolometric detector at present. Tin becomes superconducting below 3.7 K and is nonmagnetic. Thus, the specific heat of Sn is very low, having only lattice contributions at temperatures below 100 mK and can be used as a bolometric detector. On the other hand in most Nd compounds, there is a magnetic moment due to 4f electrons and the specific heat is much higher at low temperatures. This makes the Nd compounds looked at hitherto unsuitable for bolometry. At TIFR, some efforts have been already initiated to make a small cryogenic bolometer using natural Sn. The details of this effort are given in the article by **V. Nanal**.

The theoretical feasibility of ^{124}Sn as a $\beta\beta$ decay candidate was discussed by P. K. Rath. V. M. Datar, who was the coordinator of Nd-150 group after a detailed discussion concluded that ^{150}Nd is unsuitable for bolometric purpose. The summary talk was given by late Prof. C. V. K. Baba and he concluded that in fact, ^{124}Sn will be an appropriate $\beta\beta$ decay candidate to investigate 0ν $\beta\beta$ decay. During discussions, R. G. Pillay mentioned that there are some problems with "thermal links" in the case of Sn bolometer and A. Ray came forward to investigate the possibility of finding some suitable material for the thermal links. Finally, it was proposed that the feasibility study of ^{124}Sn bolometer should be initiated at national level and proposals to DST and DAE-BRNS will be submitted under multi-institutional collaboration.

3.3. NDBD07

A one-day workshop on 'Neutrinoless Double Beta Decay' (NDBD) was held at the Physical Research Laboratory (PRL), Ahmedabad on February 5, 2007 to discuss the status of the progress made by different groups that were formed in the IIT Kharagpur DBD05 workshop and in particular the progress made in the feasibility study of ^{124}Sn experiment as decided in the Lucknow NDBD05 workshop. In the PRL workshop, there were some review talks and also talks on the progress made by the groups. At the end there was an extensive discussion session, where the future plan of actions were debated and three recommendations were made. They are as follows:

1. For good progress in DBD research, human resource generation is of utmost importance. In this connection it was recommended that some MOU should be signed between the different institutes, involved in NDBD research activity, where the provision for manpower exchange should be given priority.
2. At present the experimental work towards R & D of cryogenic bolometer is being carried out at TIFR, where the process to setup a low temperature facility using a refurbished dilution refrigerator is underway. It is highly desirable to augment these efforts by initiating the cryogenic (mK) experimentation at other institutes. In this connection, it was felt strongly that the group in the Eastern region should also have a dilution refrigerator so that they can equally participate in the experimental programme.
3. Proceedings of the workshop should be published so that it will serve as a reference for future NDBD and related research programmes in India.

In addition: (i) it is felt by some participants and also by Prof. D.P. Roy (as discussed in this foreword) that the possibility of dark matter search should also be explored by the same double beta decay group as low background facilities that are an essential part of NDBD experiments are also being used everywhere for dark matter studies; (ii) in many informal discussions during the meeting, it is felt that there should be more consolidated effort in nuclear model calculations and particle physics studies (these and numerical simulation studies require dedicated computational facilities) and it is important to have frequent group meetings of members from different institutes/universities associated with this programme. The items (i) and (ii) need further discussions and they may be taken up in the next NDBD meeting planned at TIFR (Mumbai).

4. Present Status and Future Perspective

One proposal to DST in March, 2006 from IIT, Kharagpur and the other to DAE-BRNS in June, 2006 have been submitted as multi-institutional collaboration. In the DST project, the proposal is to design a prototype of natural Sn bolometer of mass 0.5-1 kg. At present, this would be housed at TIFR using a refurbished ^{3}He-^{4}He dilution refrigerator with a cooling power of 20 mW at 30 mK. There is a need to add high resistance measurement setups for mK thermometry. For this purpose, a high precision nanovoltmeter, current/voltage sources and measurement devices are needed. The detector will be qualified with α and β sources. The simulation of background and detector will be taken up by P. K. Raina and his group (details can be found in the article by **A. K. Singh**). The nuclear structure aspect will be studied by P. K. Rath. Presently, the DST proposal has been accepted. In DAE-BRNS project, requirement of materials for a prototype experimental setup has been proposed, which is under process. Finally, the option for the enrichment of ^{124}Sn will be explored.

The option for initiating collaboration with internationally active groups is also being explored. The purpose of such collaboration is to gain experience in available technical knowhow, carry out some R and D at the moment and get into collaboration, if possible at appropriate time in future with a specific multi-national group. We are already in touch with Prof. Rita Bernabei, the spokesperson of DAMA (Dark Matter) group, Gran Sasso, who is also a world leader in the study of e^+DBD and possible help as well as further collaborations are under discussion. Similarly, possibilities of collaboration with existing or proposed experiments of SNOLAB regarding double beta decay are being explored.

References

[1] B. Pontecorvo, Zh. Eksp. Teor. Fiz. **33**, 549 (1957); Z. Maki, M. Nakagawa and S. Sakata, Prog. Theor. Phys. **28**, 870 (1962).

[2] Y. Fukuda *et al.*, Phys. Rev. Lett. **81**, 1562 (1998); Y. Fukuda *et al.*, Phys. Rev. Lett. **82**, 1810 (1999); ibid **82**, 2430 (1999); ibid **82**, 2644 (1999).

[3] Q.R. Ahmed *et al.*, Phys. Rev. Lett., **87**, 071301 (2001), Q.R. Ahmed *et al.*, Phys. Rev. Lett., **89**, 011301 (2002); S.N. Ahmed *et al.*, Phys. Rev. Lett. **92**, 181301 (2004).

[4] M.H. Ahn *et al.*, Phys. Rev. Lett. **90**, 041801 (2002).

[5] K. Eguchi *et al.*, Phys. Rev. Lett. **90**, 021802 (2003).

[6] M. Apollonio *et al.*, Eur. Phys. J. C **27**, 331 (2003).

[7] F. Boehm *et al.*, Phys. Rev. D **62**, 072002 (2000).

[8] M. Goeppert-Mayer, Phys. Rev. **48**, 512 (1935).

[9] W. Furry, Phys. Rev. **56**, 1184 (1939).

[10] E. Majorana, Nuovo Cim. **5**, 171 (1937).

[11] G. Racah, Nuovo Cim. **4**, 322 (1937).

[12] V.I. Tretyak and Y.G.Zdesenko, Atomic data and Nuclear data Tables, **80**, 83 (2002), ibid **61**, 43 (1995).

[13] Y. Chikashige, R. N. Mohapatra and R. D. Peccei, Phys. Rev. Lett. **45**, 1926 (1980); Phys. Lett. **98B**, 265 (1981).

[14] C. S. Aulakh and R. N. Mohapatra, Phys. Lett. **119B**, 136 (1982).

[15] P. Bamert, C.P. Burgess and R.N. Mohapatra, Nucl. Phys. B **449**, 25 (1995).

[16] H.V. Klapdor-Kleingrothaus *et al.*, Mod. Phys. Lett. A **16**, 2409 (2001); hep-ph/0103062.

[17] H.V. Klapdor-Kleingrothaus, I.V. Krivoshina, A. Dietz and O. Chkvorets, Phys. Lett. B **586**, 198 (2004).

[18] http://www.imsc.res.in/~ino/OpenReports/INOReport.pdf

NEUTRINOLESS DOUBLE BETA DECAY
Editors: V.K.B. Kota and U. Sarkar

IMPORTANCE OF NEUTRINOLESS DOUBLE BETA DECAY

Utpal Sarkar

Physical Research Laboratory, Ahmedabad - 380 009, India

Abstract

A natural explanation for the smallness of the neutrino mass requires them to be Majorana particles violating lepton number by two units. Since lepton number violation can have several interesting consequences in particle physics and cosmology, it is of utmost importance to find out if there is lepton number violation in nature and what is its magnitude. The neutrinoless double beta decay experiment can answer these questions: if there is lepton number violation and if neutrinos are Majorana particles. In addition, the magnitude of neutrinoless double beta decay will constrain any other lepton number violating processes. This lepton number violation may also be relatd to the matter-antimatter asymmetry of the universe, dark matter and cosmological constant.

1. Introduction

Unlike any other particles, the existence of neutrinos was postulated to explain energy-momentum conservation in the beta decay. For our understanding of the macroscopic world, the existence of neutrinos are not required. All the phenomena we see around us can be explained by the electromagnetic and the gravitational interactions. In addition the strong interaction is required to explain how the positively charged protons could stay together inside the nucleus. Only the beta decay involved the neutrinos and it interacts very weakly.

The neutrinos are highly puzzling and it took more than sixty years to find out that it has a very small mass. The atmospheric and the solar neutrinos [1, 2], combined with the Laboratory experiments like KamLAND [3] have now established that the mass-squared difference between any two of the three neutrinos are non-vanishing:

$$\begin{aligned} \Delta m^2_{atm} &= 2.1 \times 10^{-3}\ \mathrm{eV}^2 \quad \text{with} \sin^2 2\theta_{atm} > 0.92 \\ \Delta m^2_{sol} &= 7.9 \times 10^{-5}\ \mathrm{eV}^2 \quad \text{with} \tan^2 \theta_{sol} 0.4 \pm 0.1\,, \end{aligned} \tag{1}$$

where θ_{atm} is the mixing angle between ν_μ and ν_τ and θ_{sol} is the mixing angle between ν_e and one of the other two physical states, which is an admixture of the states ν_μ and ν_τ. The absolute mass of the neutrinos have not yet been determined, although there is an upper bound on the sum over all neutrino masses from cosmology [4]:

$$\sum_{i=e,\mu,\tau} m_{\nu_i} \leq 0.69\ \mathrm{eV}. \tag{2}$$

The neutrinoless double beta decay also gives an upper bound on the absolute mass of the neutrinos [5], but this bound is not valid if the neutrinos are Dirac particles. We shall come back to this discussion later.

2. Dirac and Majorana Neutrinos

The smallness of the neutrino mass can be naturally explained in the standard model, if the neutrinos are Majorana particles. A Majorana particle has the property that it is its own antiparticle. The main difference between a Majorana particle and a Dirac particle lies in their mass terms:

$$\begin{aligned} \text{Majorana particle} &: \quad M_{maj}\ \Psi\,\Psi \\ \text{Dirac particle} &: \quad M_{dir}\ \bar{\Psi}\,\Psi. \end{aligned}$$

All charged fermions are Dirac particles, since the Majorana mass terms do not conserve any charge. Only the neutrinos can have either Dirac or Majorana masses. Since the neutrinos carry lepton number, lepton number will be violated if neutrinos are Majorana particles. In the standard model, lepton number is exactly conserved and we have not observed any lepton number violation in nature so far.

Let us now write the mass terms in chiral notation. We define the left-handed and right-handed particles as:

$$\psi_L = \frac{1-\gamma_5}{2}\,\psi \quad \text{and} \quad \psi_R = \frac{1+\gamma_5}{2}\,\psi\,.$$

We now denote the parity transformation $[(\vec{x},t) \leftrightarrow (-\vec{x},t)]$ by $\mathscr{P}$, charge conjugation [particle $\leftrightarrow$ antiparticle] by $\mathscr{C}$ and time reversal $[(\vec{x},t) \leftrightarrow (\vec{x},-t)]$ as $\mathscr{T}$. The chiral fields transform under $\mathscr{C}$, $\mathscr{P}$ and $\mathscr{C}\mathscr{P}$ as:

$$\begin{array}{cccccc} \psi_L & \xleftrightarrow{\mathscr{P}} & \psi_R & (\psi^c)_L & \xleftrightarrow{\mathscr{P}} & (\psi^c)_R \\ \psi_L & \xleftrightarrow{\mathscr{C}} & \psi^c{}_L & \psi_R & \xleftrightarrow{\mathscr{C}} & \psi^c{}_R \\ \psi_L & \xleftrightarrow{\mathscr{C}\mathscr{P}} & \psi^c{}_R & \psi_R & \xleftrightarrow{\mathscr{C}\mathscr{P}} & \psi^c{}_L, \end{array}$$

where the charge conjugation is defined as $\psi^c = C\bar{\psi}^T = C\gamma_0\psi^*$, with $C = -i\gamma_2\gamma_0$ so that $(\psi^c)_L = (\psi_R)^c = \frac{1}{2}(1-\gamma_5)\psi^c$ and $\psi^c{}_R = (\psi_L)^c = \frac{1}{2}(1+\gamma_5)\psi^c$. The $\mathscr{C}\mathscr{P}\mathscr{T}$ theorem ensures that the $\mathscr{C}\mathscr{P}$ conjugate states of any field must always be present. So, any theory can have the left-handed fields ψ_L and its $\mathscr{C}\mathscr{P}$ conjugate state $\psi^c{}_R$. The mass term requires the field ψ_R and its $\mathscr{C}\mathscr{P}$ conjugate state $\psi^c{}_L$.

Denoting a neutrino by ψ, the most general mass term can be written as

$$\begin{aligned} \mathscr{L}_{mass} &= -\frac{1}{2}m_L\overline{\psi_L{}^c}\psi_L - \frac{1}{2}m_R\overline{\psi_R{}^c}\psi_R - m_D\bar{\psi}_R\psi_L + h.c. \\ &= \frac{1}{2}m_L\psi_L^TC^{-1}\psi_L + \frac{1}{2}m_R\psi_R^TC^{-1}\psi_R + m_D\psi_L^TC^{-1}\psi^c{}_L + h.c. \\ &= \frac{1}{2}(\psi \quad \psi^c)_L^TC^{-1}\begin{pmatrix} m_L & m_D \\ m_D & m_R \end{pmatrix}\begin{pmatrix} \psi \\ \psi^c \end{pmatrix}_L = \frac{1}{2}\Psi_L^TC^{-1}M\Psi_L, \end{aligned} \quad (3)$$

where $\Psi_L^T = (\psi \quad \psi^c)_L^T$ and M is a 2×2 symmetric mass matrix $M = M^T$.

This general mass term contains most of the information required for an understanding of the Dirac and Majorana masses of the neutrinos. Any theory can have only the left-handed neutrinos and its $\mathscr{C}\mathscr{P}$ conjugate state, but no right-handed neutrinos. In this case the neutrino could be massless or can have a Majorana mass m_L. When both the left-handed and right-handed neutrinos are present, several possibilities can emerge.

- the neutrinos are massless, so there are two Weyl spinors.
- $m_L = m_R = 0$, so that the left-handed and the right-handed neutrinos combine to form a Dirac neutrino.
- $m_L = m_R \neq 0$ and $m_D = 0$, so that there are two Majorana neutrinos and the physical states are ψ_L and ψ_R with masses m_L and m_R.
- m_L or m_R or both are non-vanishing, and $m_D \neq 0$. In this case also it corresponds to two Majorana neutrinos and the physical states are admixtures of the states ψ_L and ψ_R with masses obtained by diagonalising the mass matrix.

The difference between a Majorana and a Dirac particle is that, for a Dirac particle the mass term takes a left-handed particle (ψ_L) to a right-handed particle (ψ_R), while for a Majorana particle the mass term takes a left-handed particle (ψ_L) to a right-handed antiparticle ($\psi^c{}_R$, which is a $\mathscr{C}\mathscr{P}$ conjugate of a left-handed particle) or takes a right-handed particle (ψ_R) to a left-handed antiparticle ($\psi^c{}_L$). Another important difference between a Dirac and Majorana particles is the conservation of charges. If neutrinos are Majorana particles, then the mass term violates lepton number by two units.

A direct consequence of the lepton number violation is neutrinoless double beta decay. In some even-even nuclei ordinary beta decay is forbidden, although double beta decay (with and without two neutrinos) could still be allowed

$$n+n \rightarrow p+p+e^-+e^-+\nu_e+\nu_e \quad (4)$$
$$n+n \rightarrow p+p+e^-+e^-. \quad (5)$$

The $2\nu\beta\beta$ decay (equation 4) has been observed, in which the total kinetic energy of the two electrons is less than the total energy available, while for the neutrinoless double beta ($0\nu\beta\beta$) decay the total kinetic energy of the two electrons is same as the Q value of the decay. This makes it possible to distinguish these two processes.

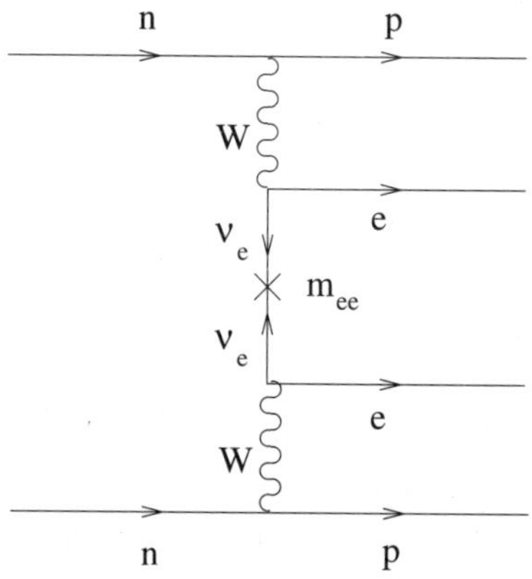

Figure 1. Majorana mass of the neutrinos allowing neutrinoless double beta decay.

When the neutrinos are Majorana particles there will be total lepton number L violation, which will allow neutrinoless double beta decay through the diagram given in figure 1. Here the neutrinos are virtual particles in the intermediate state, so the neutrino masses and mixing enter into the neutrino propagator. The half-life of the neutrinoless double beta decay thus depends on the effective neutrino mass that enters in the amplitude. The Heidelberg-Moscow $0\nu\beta\beta$ decay experiment looked for the decay mode

$$^{76}Ge \rightarrow {}^{76}Se + 2\,e^-$$

with their high resolution Ge detectors and given a strong bound on the effective mass of the neutrinos [5], $m_{ee} < 0.2$ eV. They also reported a few events for the $0\nu\beta\beta$ decay, which is yet to be confirmed.

3. Lepton Number Violation

In the standard model, there are three left-handed neutrinos ν_{iL}, $i = e, \mu, \tau$ that transform under $SU(3)_c \times SU(2)_L \times U(1)_Y$ as $(1, 2, -1)$. Thus the Majorana mass term is not allowed. Since there are no right-handed neutrinos, the Dirac mass of the neutrinos are also not allowed. Thus neutrinos are massless in the standard model. However, a natural explanation for the observed tiny neutrino mass is possible in some extensions of the standard model. A general approach to understand this is to consider the most general dimension-5 effective lepton-number violating operator in the standard model that can contribute to the Majorana masses of the neutrinos [6]

$$\mathscr{L}_{Maj} = \Lambda^{-1}(\nu\phi^{\circ} - e\phi^{+})^2. \tag{6}$$

Here Λ is some lepton-number violating heavy scale in the theory and ϕ is the Higgs doublet scalar. The electroweak symmetry breaking ($\langle\phi\rangle \neq 0$) then induce a Majorana mass for the neutrinos

$$\mathscr{L}_{Maj} = m_\nu \nu_{iL}^T C^{-1} \nu_{iL}, \tag{7}$$

where $m_\nu = v^2/\Lambda$. A large lepton number violating scale Λ can thus explain naturally why m_ν is much smaller than the charged fermion masses. This also suggests that a Majorana mass of the neutrinos is more natural than a Dirac mass.

The simplest extension of the standard model in which the effective operator 6 may be realized requires either right-handed neutrinos or triplet Higgs scalars. In models with the right-handed neutrinos, one extends the standard model with three right-handed neutrinos $N_{iR}, i = 1,2,3$, which are singlets under the standard model. The mass terms for the neutrinos are now given by

$$\begin{aligned} \mathscr{L}_{mass} &= m_D\, \nu_L\, N^c{}_L + M_R\, N^c{}_L\, N^c{}_L + H.c. \\ &= (\nu \quad N^c)_L \begin{pmatrix} 0 & m_D \\ m_D^T & M_R \end{pmatrix} \begin{pmatrix} \nu \\ N^c \end{pmatrix}_L . \end{aligned} \tag{8}$$

Here the 3×3 mass matrix m_D originates from the standard model Higgs vacuum expectation value, so it is of the order of charged lepton masses. But the Majorana mass of the right-handed neutrinos, which is the lepton number violating scale in the theory, could be very large: $M_R \sim 10^{10}$ GeV. Thus assuming $m_D \ll M$, the eigenvalues of this mass matrix then become,

$$m_1 = -\frac{m_D^2}{M_R} \quad \text{and} \quad m_2 = M_R. \tag{9}$$

We then get a light neutrino with mass $m_1 \sim 0.1$ eV, which is mostly the left-handed neutrino with a small mixing $\tan\theta = \frac{2M}{m_D}$ with the right-handed neutrino. This is also known as the see-saw mechanism of neutrino masses [7]. This small neutrino mass will contribute to the neutrinoless double beta decay. Thus the neutrinoless double beta decay can, in principle, determine the absolute mass scale of the neutrinos.

We shall now consider another equivalent realization of the effective operator (6), where the standard model is extended to include a triplet Higgs scalar ξ, which transforms under $SU(3)_c \times SU(2)_L \times U(1)_Y$ as $[1,3,+1]$ [8, 9]. Its couplings to the leptons and the standard model Higgs doublet ϕ break lepton number,

$$\mathscr{L}_{Yuk} = f_{ij}\, \xi\, \ell_i \ell_j + \mu\, \xi^\dagger\, \phi\phi. \tag{10}$$

We consider [9] the possibility $\mu \neq 0$ ($\mu = 0$ models [8] are ruled out by LEP data). The neutral component of ξ will acquire a induced *vev* during the electroweak symmetry breaking $u = \frac{-\mu v^2}{M^2}$, where M is the mass of the triplet Higgs ξ. The lepton number is broken explicitly at a very high scale $M \sim \mu$. So, there are no Goldstone bosons corresponding to the broken lepton number symmetry. The mass of the left-handed neutrinos are then given by

$$m^{\nu}_{ij} = f_{ij} u = -f_{ij} \frac{\mu v^2}{M^2}, \tag{11}$$

which is of the order of $\sim$ eV. The neutrino mass matrix is now directly proportional to the Yukawa couplings f_{ij}. The absolute mass scale can be determined by the neutrinoless double beta decay.

The smallness of the neutrino mass can thus be naturally explained if neutrinos are Majorana particles and lepton number is violated by two units. The Majorana nature of the neutrinos can be confirmed by the neutrinoless double beta decay. In fact, any lepton number violating processes can contribute to the neutrinoless double beta decay. So, all lepton number violating processes are constrained by the neutrinoless double beta decay [10]. In left-right symmetric models the right-handed charged gauge boson mass and the right-handed neutrino mass could be constrained by the present bound on the $0\nu\beta\beta$ decay. The inverse beta decay are also strongly constrained. The leptoquarks, diquarks and other exotic scalar bilinears that couples to two fermions of the standard model are also constrained by the $0\nu\beta\beta$ decay. In supersymmetric models all the R-parity violating and lepton number violating couplings are strongly constrained by the $0\nu\beta\beta$ decay. Even in R-parity conserving supersymmetric models, there could be lepton number violation originating from the soft terms, which are also constrained by the $0\nu\beta\beta$ decay. The compositeness scale for some models with composite particles are also constrained by the $0\nu\beta\beta$ decay. Some of these constraints and the consequence of $0\nu\beta\beta$ decay in colliders will be reviewed in another article by Prof. S.D. Rindani in this proceedings. We shall now proceed to discuss some cosmological consequences of the lepton number violation.

4. Matter-Antimatter Asymmetry

Our universe is composed mainly of matter and very little antimatter. This matter dominance requires an explanation, since a natural choice would be to start with a universe that is neutral with respect to any conserved charges like baryon or lepton numbers. At present the most popular explanation of this matter dominance in the universe originates from the lepton number violation that is required for the Majorana neutrino masses. This is known as leptogenesis. The present limit on the amount of lepton number violation coming from the neutrinoless double beta decay is just enough to explain this matter dominance and this predictability makes this scenario more appealing. To establish this connection between the neutrinoless double beta decay and leptogenesis, we shall discuss leptogenesis in the see-saw model and the triplet Higgs scalar model discussed in the previous section.

The generation of the baryon asymmetry of the universe requires three ingredients: *i*) Baryon number (B) violation, *ii*) $\mathscr{CP}$ violation, and *iii*) departure of the B-violating interactions from equilibrium. On the other hand if lepton number (L) is violated satisfying all the three conditions, then that will generate a lepton asymmetry of the universe. In the standard model, both B and L are global symmetries, but $(B+L)$

is broken by quantum effects arising from anomalous triangle loop diagrams. These anomaly induced $B+L$ violating processes are suppressed by the quantum tunnelling probability. But at finite temperature, during the period $10^2 < T < 10^{12}$ GeV, these interactions become strong in the presence of some static topological field configuration called the sphalerons [11]. As a result any existing L asymmetry of the universe will get converted to the required baryon asymmetry of the universe, before the electroweak phase transition [12, 13].

In the see-saw mechanism of neutrino masses, the Majorana masses of the heavy right-handed neutrinos violate lepton number. $\mathscr{CP}$ violation comes from the complex Yukawa couplings and interference of tree level decays with one-loop diagrams. These interactions can also satisfy the out-of-equilibrium condition. Thus the decays of the right-handed neutrinos into a lepton ($N_{Ri} \rightarrow \ell_{jL} + \bar{\phi}$) and also an antilepton ($N_{Ri} \rightarrow {\ell_{jL}}^c + \phi$) can generate a lepton asymmetry of the universe, which then get converted to a baryon asymmetry of the universe in the presence of the sphalerons.

The $\mathscr{CP}$ violation comes from an interference of the tree level decays of the right-handed neutrinos and the one loop diagrams:

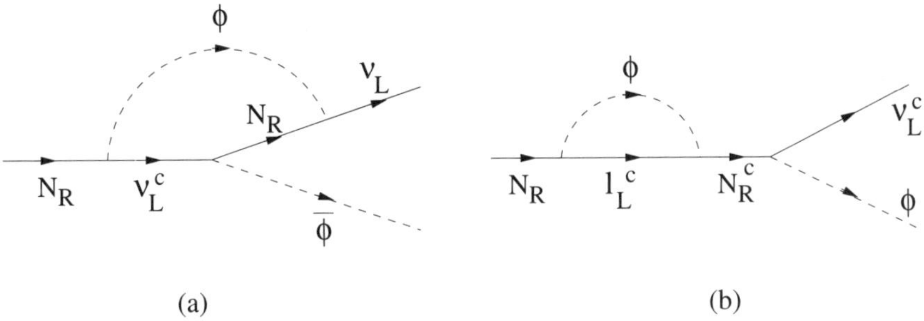

Figure 2. One loop (a) vertex and (b) self energy diagrams, which interferes with the tree level right-handed neutrino decays to produce CP violation.

(i) vertex diagram [12, 14] of figure 2a, which is similar to the CP violation coming from the penguin diagram in $K-$decays.

(ii) self energy diagram [15] of figure 2b, which is similar to the CP violation in $K-\bar{K}$ oscillation, entering in the mass matrix of the heavy Majorana neutrinos.

The interference gives an asymmetry

$$\delta = \frac{\Gamma(N \rightarrow \ell\phi^\dagger) - \Gamma(N \rightarrow \ell^c\phi)}{\Gamma(N \rightarrow \ell\phi^\dagger) + \Gamma(N \rightarrow \ell^c\phi)}, \tag{12}$$

which, when satisfies the out-of-equilibrium condition:

$$\Gamma(N \rightarrow \ell\bar{\phi}) < 1.7\sqrt{g_*}\frac{T^2}{M_P} \qquad \text{at } T = M_N, \tag{13}$$

can generate the required lepton asymmetry of the universe. Here the right-hand-side correspond to the expansion rate of the universe and M_P is the Planck scale. The lepton asymmetry thus generated is same as the $(B-L)$ asymmetry of the universe, since there is no primordial baryon asymmetry at this time. The sphaleron interactions now convert this $(B-L)$ asymmetry to a baryon asymmetry of the universe.

The amount of lepton asymmetry depends on the Yukawa couplings and the out-of-equilibrium condition also depends on the Yukawa couplings. Both these conditions can be satisfied for certain range of parameters, which implies a neutrino mass of $m_\nu <$ 0.2 eV [16]. Although this limit is consistent with the upper bound on the neutrinoless double beta decay, the reported events for the neutrinoless double beta decay is not consistent with this limit [5]. Thus determination of the neutrinoless double beta decay half-life will tell us if the simplest version of leptogenesis is possible.

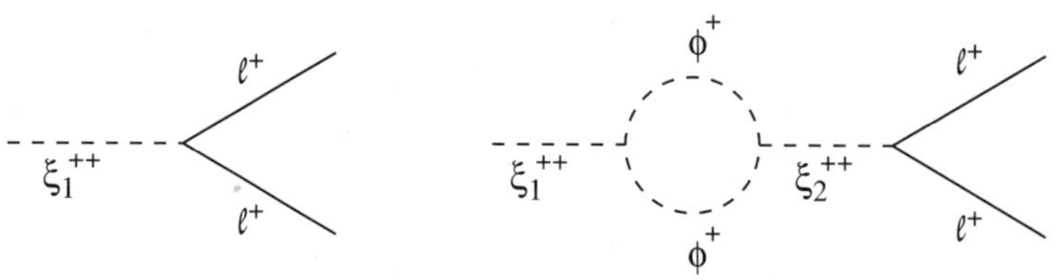

Figure 3. The decay of $\xi_1^{++} \to l^+ l^+$ at tree level and in one-loop order, whose interference gives $\mathscr{CP}$ violation.

The triplet Higgs mechanism of neutrino masses [9] can also allow leptogenesis. The decays of the triplet Higgs scalars $\xi_a, a = 1,2$, two scalars are required for $\mathscr{CP}$ violation) violate lepton number

$$\xi_a^{++} \to \begin{cases} l_i^+ l_j^+ & (L=-2) \\ \phi^+\phi^+ & (L=0) \end{cases} \tag{14}$$

$\mathscr{CP}$ violation from the interference of the tree-level decays and the self energy diagrams of figure 3. The rate of $\xi_b \to \xi_a$ no longer remains to be the same as $\xi_b^* \to \xi_a^*$. Since by *CPT* theorem $\xi_b^* \to \xi_a^* \equiv \xi_a \to \xi_b$, it means

$$\Gamma[\xi_a \to \xi_b] \neq \Gamma[\xi_b \to \xi_a]. \tag{15}$$

This is a different kind of CP violation compared to the CP violation in models with right-handed neutrinos. The lepton asymmetry is now given by,

$$\delta = \frac{\Gamma(\xi \to \ell\ell) - \Gamma(\xi^\dagger \to \ell^c\ell^c)}{\Gamma(\xi \to \ell\ell) + \Gamma(\xi^\dagger \to \ell^c\ell^c)}. \tag{16}$$

The out-of-equilibrium condition is satisfied when the triplet Higgs scalars are very heavy. In this case the required amount of lepton asymmetry do not constrain the neutrino masses.

5. Dark Matter and Dark Energy

The total matter in the universe is same as the critical density and about 25% of the matter is dark matter and about 70% of matter is in the form of dark energy. Only about 5% matter is baryonic matter, of which only a fraction is visible. There are several dark matter candidates, including the lightest supersymmetric particle, which is stable and very weakly interacting. One class of dark matter candidate is related to the neutrino masses. If some discrete symmetry forbids the Yukawa coupling relating the left-handed and the right-handed neutrinos, there could be a second Higgs doublet scalar that does not acquire any *vev* or interact with the charged fermions and remain inert. The lightest of these inert particles (LIP) then could be a dark matter candidate [17].

The problem with dark energy is that the large symmetry breaking scales in particle physics would contribute orders of magnitude large dark energy, while observations indicate that the dark energy is comparable to the dark matter content of the universe. A natural solution is thus to consider a scenario in which the dark energy varies with time starting from a very high value in the early universe. In a popular model, the mass density of a scalar field, called the quintessence, gives the dark energy [18, 19]. The potential of the quintessence field ensures that the decrease of the dark energy is slower than the mass densities of matter and radiation, so that the nucleosynthesis predictions are not altered. Recently it has been pointed out that a varying neutrino mass scenario can account for the dark energy of the universe [20, 21]. The variation of the neutrino mass can originate from some scalar field, which could be a pseudo-Nambu-Goldstone boson in the neutrino sector [22].

References

[1] *Super-Kamiokande Collaboration:* Y. Fukuda, et al., Phys. Lett. **B 433**, 9 (1998); Phys. Rev. Lett. **81**, 1562 (1998); Phys. Rev. Lett. **82**, 2644 (1999).

[2] R. Davis, Jr, Phys. Rev. Lett. **12**, 303 (1964); Super-Kamiokande Collaboration: Y. Fukuda *et al*, Phys. Rev. Lett. **81**, 1158 (1998); Phys. Rev. Lett. **86**, 5656 (2001); SNO Collaboration: Q.R. Ahmad, *et. al.*, Phys. Rev. Lett. **87**, 071301 (2001).

[3] *KamLAND Collaboration:* T. Araki, et al., Phys. Rev. Lett.**94**, 081801 (2005).

[4] D.N. Spergel, et al., Astrophys. J. Suppl. **148**, 175 (2003).

[5] *Heidelberg–Moscow Collaboration:* H.V. Klapdor-Kleingrothaus et al., Phys. Rev. **D 55**, 54 (1997); Phys. Lett. **B 407**, 219 (1997); Phys. Rev. Lett. **83**, 41 (1999); Eur. Phys. J. **A 12**, 147 (2001); Mod. Phys. Lett. **A 16**, 2409 (2001).

[6] S. Weinberg, Phys. Rev. Lett. **43**, 1566 (1979); F. Wilczek and A. Zee, Phys. Rev. Lett. **43**, 1571 (1979); E. Ma, Phys. Rev. Lett. **81**, 1171 (1998);

[7] P. Minkowski, Phys. Lett. **B 67**, 421 (1977); M. Gell-Mann, P. Ramond and R. Slansky, in *Supergravity, Proc of the Workshop, Stony Brook, New York, 1979*, ed. by P. van Nieuwenhuizen and D. Freedman (North-Holland, Amsterdam) 1979; T. Yanagida, in *Proc of the Workshop on Unified Theories and Baryon Number in the Universe*, Tsukuba, Japan, edited by A. Sawada and A. Sugamoto (KEK Report No. 79-18, Tsukuba) 1979; R.N. Mohapatra and G. Senjanovic, Phys. Rev. Lett. **44**, 912 (1980).

[8] J. Schechter and J.W.F. Valle, Phys. Rev. **D 22**, 2227 (1980); R. N. Mohapatra and R. E. Marshak, Proceedings of the Orbis Scientiae, January, 1980, p. 277 (Plenum Press, ed. B. Korsonoglu et al); T. P. Cheng and L. F. Li, Phys. Rev. **D 22**, 2860 (1980); G.B. Gelmini and M. Roncadelli, Phys. Lett. **B 99**, 411 (1981); C. Wetterich, Nucl. Phys. **B 187**, 343 (1981); G. Lazarides, Q. Shafi, C. Wetterich, Nucl. Phys. **B 181**, 287 (1981);

[9] E. Ma and U. Sarkar, Phys. Rev. Lett. **80**, 5716 (1998); E. Ma, T. Hambye, U. Sarkar, Nucl. Phys. **B 602**, 23 (2001); G. Lazarides and Q. Shafi Phys. Rev. **D 58**, 071702 (1998); E. Ma, M. Raidal, U. Sarkar, Phys. Rev. Lett. **85**, 3769 (2000); Nucl. Phys. **B 615**, 313 (2001); W. Grimus, R. Pfeiffer, T. Schwetz, Eur. Phys. J. **C 13**, 125 (2000).

[10] H.V. Klapdor-Kleingrothaus, H. Päs and A.Yu. Smirnov, Phys. Rev. **D 63**, 073005 (2001); H.V. Klapdor-Kleingrothaus and U. Sarkar, Mod. Phys. Lett. **A 16**, 2469 (2001); Proc. Ind. Nat. Sc. Acad., **70**, 251 (2004).

[11] V. Kuzmin, V. Rubakov and M. Shaposhnikov, Phys. Lett. **B 155** (1985) 36.

[12] M. Fukugita and T. Yanagida, Phys. Lett. **B 174**, 45 (1986).

[13] S.Yu. Khlebnikov and M. Shaposhnikov, Nucl. Phys. **B 308**, 885 (1988); J.A. Harvey and M.S. Turner, Phys. Rev. **D 42**, 3344 (1990).

[14] P. Langacker, R.D. Peccei and T. Yanagida, Mod. Phys. Lett. **A 1**, 541 (1986); M.A. Luty, Phys. Rev. **D 45**, 445 (1992); R.N. Mohapatra and X. Zhang, Phys. Rev. **D 45**, 5331 (1992); H. Murayama, H. Suzuki, T. Yanagida and J. Yokoyama, Phys. Rev. Lett. **70**, 1912 (1993); A. Acker, H. Kikuchi, E. Ma and U. Sarkar, Phys. Rev. **D 48**, 5006 (1993); P.J. O'Donnell and U. Sarkar, Phys. Rev. **D 49**, 2118 (1994); W. Buchmüller and M. Plümacher, Phys. Lett. **B 389**, 73 (1996); A. Ganguly, J.C. Parikh and U. Sarkar, Phys. Lett. **B 385**, 175 (1996); J. Faridani, S. Lola, P.J. O'Donnell and U. Sarkar, Z. Phys. **C 7**, 543 (1999).

[15] M. Flanz, E.A. Paschos, and U. Sarkar, Phys. Lett. **B 345**, 248 (1995); M. Flanz, E.A. Paschos, U. Sarkar and J. Weiss, Phys. Lett. **B 389**, 693 (1996); L. Covi, E. Roulet and F. Vissani, Phys. Lett. **B 384**, 169 (1996); A. Pilaftsis, Phys. Rev. **D 56**, 5431 (1997); W. Buchmüller and M. Plümacher, Phys. Lett. **B 431**, 354 (1998); T. Hambye, J. March-Russell and S.M. West, JHEP **0407**, 070 (2004); A. Pilaftsis and T.E.J. Underwood, Nucl. Phys. **B 692**, 303 (2004).

[16] W. Buchmüller, P. Di Bari and M. Plumacher Nucl. Phys. **B 643**, 367 (2002); Phys. Lett. **B 547**, 128 (2002); Nucl. Phys. **B 665**, 445 (2003).

[17] R. Barbieri, L.J. Hall and V.S. Rychkov, Phys. Rev. **D 74**, 015007 (2006); E. Ma, Mod. Phys. Lett. **A 21**, 1777 (2006).

[18] C. Wetterich, Nucl. Phys. **B 302**, 668 (1988); B. Ratra and P.J.E. Peebles, Astrophys. Jour. **325**, L17 (1988).

[19] C.T. Hill, D.N. Schramm and J.N. Fry, Nucl. Part. Phys. **19**, 25 (1989); J.A. Freeman, C.T. Hill, A. Stebbins and I. Waga, Phys. Rev. Lett. **75**, 2077 (1995).

[20] R. Fardon, A.E. Nelson and N. Weiner, JCAP **10**, 005 (2004); P. Gu, X. Wang, X. Zhang, Phys. Rev. **D 68**, 087301 (2003.); P.Q. Hung, hep-ph/0010126.

[21] E. Ma and U. Sarkar, Phys. Lett. **B 638**, 358 (2006); P. Gu, H. He and U. Sarkar, arXiv:0704.2020 [hep-ph]; R. Takahashi and M. Tanimoto, Phys. Lett. **B633**, 675 (2006); R. Fardon, A. E. Nelson, and N. Weiner, JHEP **0603**, 042 (2006); P. Gu, H-J. He and U. Sarkar, arXiv:0705.3736 [hep-ph].

[22] R. Barbieri, L.J. Hall, S.J. Oliver, A. Strumia, Phys. Lett. **B 625**, 189 (2005); C. Hill, I. Mocioiu, E.A. Paschos and U. Sarkar, Phys. Lett. **B 651**, 188 (2007).

NEUTRINOLESS DOUBLE BETA DECAY
Editors: V.K.B. Kota and U. Sarkar

CONSEQUENCES OF LEPTON FLAVOUR VIOLATION

Saurabh D. Rindani[1]

Physical Research Laboratory, Navrangpura, Ahmedabad 380 009, India

Abstract

Observation of neutrinoless double beta decay ($0\nu\beta\beta$) would signal violation of lepton number L by two units. While the simplest origin of this violation could be neutrino Majorana mass terms, which correspond to $\Delta L = 2$, other interactions violating lepton number can contribute to $0\nu\beta\beta$. Some of these interactions and their contributions to $0\nu\beta\beta$ are discussed. Such interactions also imply the existence of Majorana mass for neutrinos via the "black box theorem". Processes other than $0\nu\beta\beta$ which can signal lepton-number violation are described. Some recent proposals for observing lepton-number violation at colliders are reviewed.

[1] **E-mail:saurabh@prl.res.in**

1. Introduction

Neutrinoless double beta decay ($0\nu\beta\beta$)

$$nn \rightarrow ppe^-e^-, \tag{1}$$

if observed, would indicate violation of lepton number by two units ($\Delta L = 2$). The most obvious mechanism which could contribute to $0\nu\beta\beta$ is one involving a possible Majorana mass of a neutrino, which violates lepton number by two units [1]. The contribution of Majorana mass of neutrino(s) to neutrinoless double beta decay is represented by the diagram in Fig. 1.

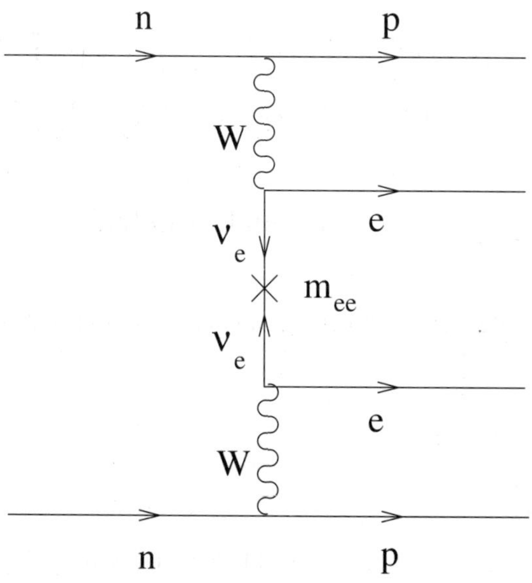

Figure 1. Diagram showing the contribution of a Majorana neutrino to neutrinoless double beta decay. The Majorana mass of the neutrino, m_{ee}, is represented by a cross.

The lifetime for $0\nu\beta\beta$ is given by

$$\left[T_{1/2}^{0\nu\beta\beta}\right]^{-1} = G^{0\nu}(Q,Z)|M^{0\nu}|^2|m_{\text{eff}}|^2, \tag{2}$$

where the $G^{0\nu}(Q,Z)$ is the phase space factor, and $M^{0\nu}$ is the nuclear matrix element for the process. The"effective Majorana" mass m_{eff} is the matrix element of the neutrino mass matrix in the basis in which the charged-lepton mass matrix is diagonal. For three generations of light neutrinos, m_{eff} can be expressed as

$$M_{ee}^{\nu} \equiv m_{\text{eff}} = c_{12}^2 c_{13}^2 m_1 + s_{12}^2 c_{13}^2 e^{i\alpha} m_2 + s_{13}^2 e^{i\beta} m_3, \tag{3}$$

where c_{ij}, s_{ij} refer to the cosine and sine of the three-generation mixing angle θ_{ij} of the neutrinos, m_i are the neutrino mass eigenvalues, and α, β are the phases occuring in the neutrino mixing matrix. Eq.(3) contains a priori seven unknowns: Three mass eigenstates, two angles and two phases. With the help of data from neutrino oscillation experiments, one can trade two mass eigenstates for the observed $\Delta m^2_{\rm Atm}$ and $\Delta m^2_\odot$ and relate the two angles to the solar ($\theta_\odot$) and reactor angle (θ_R). Though $m_{\rm eff}$ can be extracted from $0\nu\beta\beta$ experimental result, because of the unknown Majorana phases, $m_{\rm eff}$ does not permit unambiguous determination of the mass orderings of the neutrinos.

Limited information can be derived in certain cases. For example, if one can determine $m_{\rm eff} \geq 0.1$ eV, the mass pattern is degenerate. If $m_{\rm eff} \leq 0.01 - 0.02$ eV, but nonvanishing, the hierarchy is normal. For inverted hierarchy, $m_{\rm eff} > 0.01$ eV. The value of 0.01 is within reach of the next generation of experiments In principle it is possible to have normal hierarchy, but $m_{\rm eff} = 0$ because of cancellations – fine tuning or underlying symmetry.

Nevertheless, the observation of $0\nu\beta\beta$ would indicate either a tree-level Majorana mass for at least one neutrino, or some mechanism for violation of lepton number. Here we look at some processes which like $0\nu\beta\beta$ would demonstrate lepton-number violation by two units. Such processes would be constrained by the present limits on $0\nu\beta\beta$ or would have implications for neutrino masses and mixings.

2. Lepton-Number Violation

Neutrinoless double-beta decay implies violation of lepton number L by two units. This could happen because of a Majorana mass term for one or more neutrinos:

$$\mathscr{L}_{\rm mass} = -m\overline{\nu}\nu^C. \tag{4}$$

The lepton-number violation (LNV) gets carried over to the charged-lepton sector via charged-current interactions and manifests itself in a process like $0\nu\beta\beta$.

Even though $0\nu\beta\beta$ is the most sensitive process in which lepton number is violated by two units, it is not the only process. Examples of other processes which violate lepton number by two units are:

1. $\mu^- + (Z,A) \to e^+ + (Z-2,A)$. The expected branching ratio is $\leq 10^{-12}$.
2. $K^+ \to \mu^+\mu^+\pi^-$. The expected branching ratio is $\leq 3\times 10^{-9}$.
3. $\overline{\nu}_e$ emission from the sun. The expected branching ratio $\leq 10^{-4}$.

There are also proposals to look for LNV processes at colliders. Some of these are described later on.

3. Black Box Theorem

There is an interesting theorem which establishes the connection between $0\nu\beta\beta$ and neutrino Majorana masses.

It is possible that there is no explicit Majorana mass term (in lowest order), but there are other $\Delta L = 2$ interactions. Examples are, lepton-number violating couplings to new Higgs bosons, R-parity violating coupling in supersymmetric theories, etc. Whatever the source of nonzero contribution to $0\nu\beta\beta$, it implies that neutrinos are massive Majorana fermions. It is possible that some symmetries prevent a Majorana mass for neutrinos at the tree level. In these cases, a nonzero M^{ν}_{ee} is induced at some order [2]. This statement is generalized recently and is called the "black box theorem" [3]. The authors of [3] also show that if M^{ν}_{ee} is somehow made to vanish, the corresponding theory is inconsistent with oscillation data.

Thus, the presence of Majorana mass is not only sufficient for $0\nu\beta\beta$, the black box theorem implies that it is also necessary for $0\nu\beta\beta$ to occur.

4. Models for Neutrinoless Double Beta Decay

4.1. Mechanism for $0\nu\beta\beta$ Decay

$0\nu\beta\beta$ can occur because of the exchange of a light but massive Majorana neutrino. LNV interactions involving new heavy (TeV scale) particles can lead to comparable $0\nu\beta\beta$ rate.

In the absence of more information about the mechanism, M^{ν}_{ee} cannot be inferred from the $0\nu\beta\beta$ decay rate.

The amplitude with light Majorana neutrino exchange, a "short-distance" contribution, is

$$A_L \approx G_F^2 \frac{M^{\nu}_{ee}}{\langle k^2 \rangle}, \tag{5}$$

where $\langle k^2 \rangle \approx (50\,\mathrm{MeV})^2$ is the neutrino virtuality. The amplitude with heavy particle exchange, a "long-distance contribution", is

$$A_H \approx G_F^2 \frac{M_W^4}{\Lambda^5}, \tag{6}$$

where Λ is the heavy scale. $A_H/A_L \approx 1$ for $M^{\nu}_{ee} \approx 0.1 - 0.5$ eV and $\Lambda \approx 1$ TeV. The fact that the amplitudes arising from the light and heavy exchange can be comparable should be kept in mind when examining specific models.

4.2. Left-Right Symmetry

One of the popular models is the left-right symmetric model. In this model, based on the gauge group $SU(2)_L \times SU(2)_R \times U(1)$, left-handed and right-handed fermions transform as doublets under the $SU(2)_L$ and $SU(2)_R$ gauge groups. Thus, there is a right-handed doublet of neutrinos. W_L and W_R couple respectively to left-handed and right-handed fermions. Fig. 2 shows the contributions of the combinations $W_L W_L$, $W_L W_R$ and $W_R W_R$ to neutrinoless double beta decay. Of these, the first (the diagram on the extreme left) is a long-distance contribution, whereas the other two are short-distance contributions, owing to the large mass of W_R. In left-right symmetric models the mixing between the active left-handed light neutrinos with the sterile right-handed heavy ones can be estimated to be $\sum_i U_{ei} V_{ei} \sim \frac{m_D}{M_M} \sim \sqrt{\frac{m_\nu}{M_M}}$, where U and V represent the left-handed and right-handed charged-current mixing matrices. This can lead to cancellations between long-range contributions [4].

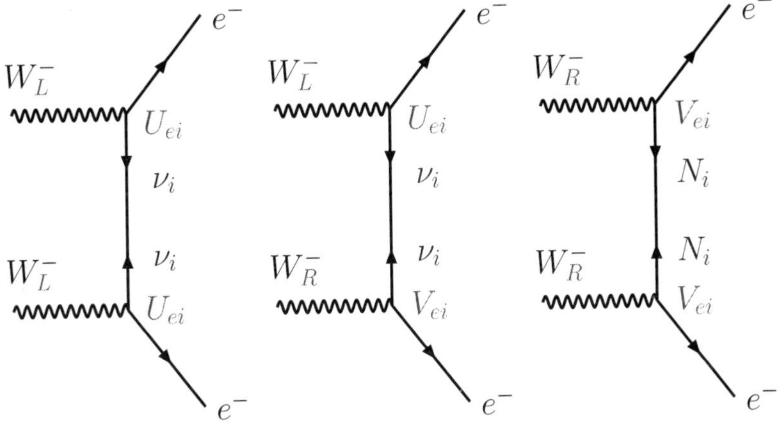

Figure 2. Contributions to neutrinoless double beta decay in the left-right symmetric model. (From [4]).

In the short range contribution, although some cancellation of terms in $\langle m_N \rangle$ might occur, no strong supression is expected. Assuming a limit on the ^{76}Ge half-live of $T_{1/2}^{0\nu\beta\beta} \geq 1.2 \cdot 10^{25}$ y, we get a limit [4]

$$m_{W_R} \geq 1.3 \left(\frac{\langle m_N \rangle}{1\text{TeV}} \right)^{-1/4} \quad \text{TeV}. \tag{7}$$

Uncertainty in this limit due to uncertainty in nuclear matrix element calculation scales as $\Delta m_{W_R} \sim (\Delta \mathscr{M})^{-1/4}$ and thus quite insensitive to the details of the nuclear model.

4.3. R-parity Violating Supersymmetry

Supersymmetry requires spin-0 partners for quarks and leptons, and spin-1/2 partners for gauge and Higgs bosons. If R parity is not assumed there can be extra lepton-number violating interactions induced by scalar partners of leptons or quarks.

A limit of $T_{1/2}^{0\nu\beta\beta} \geq 1.2 \cdot 10^{25}$ y for ^{76}Ge leads to [5, 7]

$$\lambda'_{111} \leq 3.2 \times 10^{-4} \left(\frac{m_{\tilde{q}}}{100\text{GeV}} \right)^2 \left(\frac{m_{\tilde{g}}}{100\text{GeV}} \right)^{1/2}, \tag{8}$$

where λ'_{111} is the R-parity violating coupling which couples the first generation quarks and leptons to the first generation squark, the superpartner of the quark. Such a small value of λ'_{111} generates at 1-loop level an entry in the Majorana neutrino mass matrix of $M^{\nu}_{ee} \simeq 10^{-6}$ eV only.

4.4. Other Models

Other models with exotic particles coupling to fermion bilinears exist. Spin-0 "dileptons" can couple to two leptons. The bounds from this kind of contribution are weak

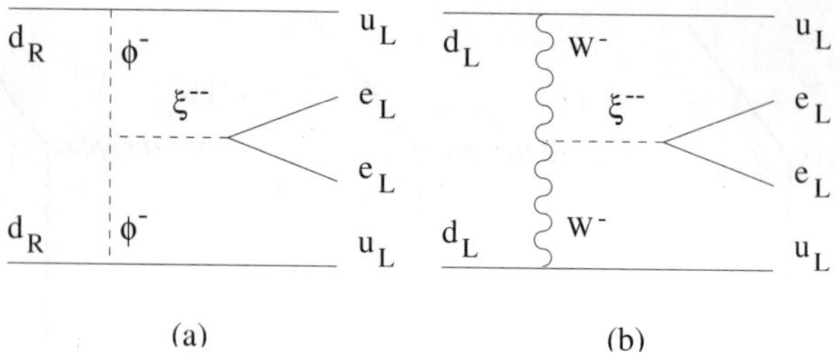

Figure 3. Diagram for neutrinoless double beta decay with dileptons.

[6]. Leptoquarks couple to bilinears of quarks and leptons. If $SU(2)$-singlet leptoquarks mix with $SU(2)$-doublet leptoquarks, they can contribute to $0\nu\beta\beta$. There can be a large enhancement, leading to a stringent limit on a combination of leptoquark couplings and mass [7].

5. Lepton-Number Violation at Accelerators

5.1. *Inverse Process* $e^-e^- \to W^-W^-$

Heavy right-handed Majorana neutrinos have a small mixing with the left-handed neutrinos, which allow these heavy neutrinos to enter in the $0\nu\beta\beta$ decay. The upper limit on the lifetime of the $0\nu\beta\beta$ decay then gives a lower limit on the mass of the heavy neutrinos with some assumption on the mixing matrix. The probability for the discovery of the inverse beta decay process $e^-e^- \to W^-W^-$ at a future linear collider (ILC), which takes place via a t-channel exchange of a heavy Majorana neutrino, could thus be constrained [8].

The present value can be achieved at the ILC when it reaches a center of mass energy of 2 TeV [8]. The discovery limit for $e^-e^- \to W^-W^-$ at the ILC for several center-of-mass energies is presented in figure 4 as a function of the heavy neutrino mass eigenstate M_i and their mixing with lighter eigenstates with masses m_α ($\alpha = e, \mu, \tau$), U_{ei}^2. This figure considers 10 events as discovery limit and assumes an unpolarized e^- beam with a luminosity of 80 $[\sqrt{s}(1\ \text{TeV})]^2$ fb^2. From the figure it is clear the for $\sqrt{s} = 500$ GeV and 1 TeV, the parameter space which produces observable $e^-e^- \to W^-W^-$ is already ruled out by the present result on the $0\nu\beta\beta$.

5.2. *The Process* $e^-e^\pm \to H^-H^\pm$

A charged-Higgs pair can be produced at an ee collider through a t-channel exchange of a heavy neutrino (N) via $e^+e^- \to H^+H^-$ and, if N is a Majorana particle, also via the lepton-number violating like-sign charged-Higgs signal $e^\pm e^\pm \to H^\pm H^\pm$. In any generic multi-Higgs model, the coupling H^+eN is $\propto \xi \times (m_N/m_W)$, where ξ is the product of the $N-e$ mixing element U_{eN} times some function of the Higgs VEV's: $\xi \equiv U_{eN} \cdot f(v_i)$. For example, $\xi = U_{eN}/\tan\beta$ in the supersymmetric version of the seesaw model, in which a right-handed singlet neutrino chiral superfield is added to the minimal supersymmetric standard model. This interaction vertex can give a striking enhancement to these charged-Higgs pair production signals [9].

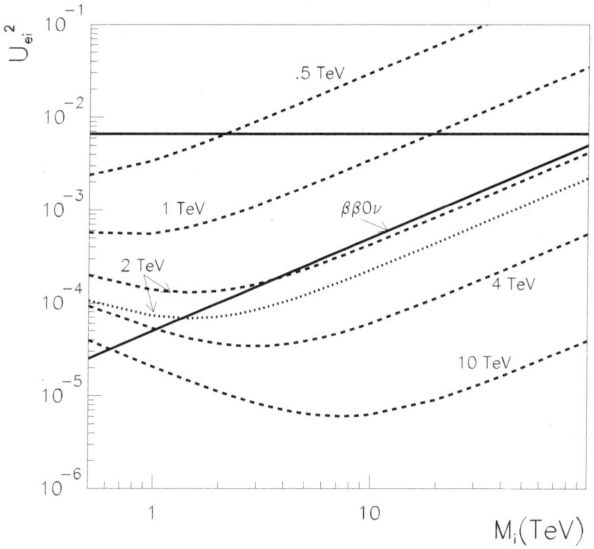

Figure 4. Discovery limit for $e^-e^- \to W^-W^-$ at the ILC is compared with the $0\nu\beta\beta$ decay result. A minimum c.m. energy of 2 TeV is seen to be required for the limits to be comparable. (From [8]).

Fig. 5, taken from [9], shows the discovery limit for the processes $e^-e^- \to H^-H^-$ and $e^-e^+ \to H^-H^+$ at a linear collider, together with limits on U_{eN}^2 from $0\nu\beta\beta$ experiments, from invisible Z decays, and from perturbative unitarity. It is found that the LNV H^-H^- signal can probe N at the ILC in the mass range 100 GeV $\lesssim m_N \lesssim 10^4$ TeV and with the effective mixing angle, ξ, in the range $10^{-4} \lesssim \xi^2 \lesssim 10^{-8}$ – well within its perturbative unitarity bound and the $\beta\beta 0\nu$ limit [9].

5.3. Like-Sign Leptons at Hadron Colliders

Han and Zhang [10] have explored the possibility of $\Delta L = 2$ like-sign dilepton production at hadron colliders as signals of Majorana neutrinos. They find significant sensitivity from resonant production of a Majorana neutrino in the mass range of 10 – 80 GeV at the current run of the Tevatron with 2 fb^{-1} integrated luminosity, and in the range of 10 – 400 GeV at the LHC with 100 fb^{-1}.

That a Majorana neutrino can be produced as a resonance if kinematically accessible was noted earlier in [11], where only $m_N > 100$ GeV for the Majorana neutrino mass was considered for the $e^{\pm}e^{\pm}$ mode. Han and Zhang extended the calculation to include an arbitrary mass m_N and general mixing for final state lepton flavors. Earlier work had demonstrated that the experimental constraint from $0\nu\beta\beta$ is too strong if the signal looked for consists of electrons. Han and Zhang therefore consider the possibility that the final state leptons are muons, which are easier to detect experimentally.

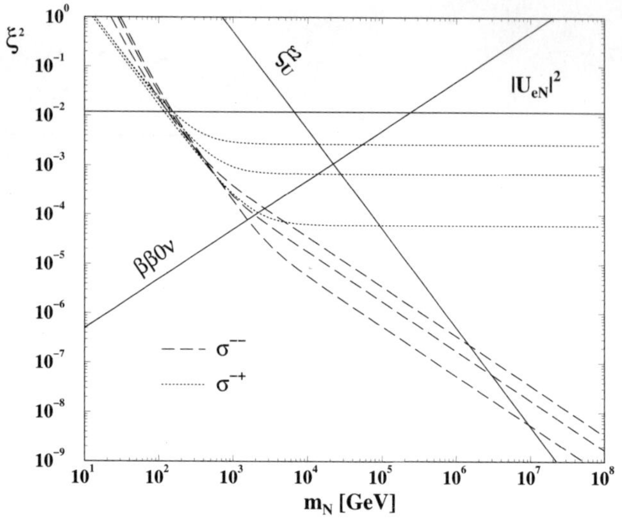

Figure 5. Discovery limit for $e^-e^- \to H^-H^-$ (long-dashed lines) and for $e^-e^+ \to H^-H^+$ (dotted lines), at an ILC with $\sqrt{s} = 500$ GeV (upper line), $\sqrt{s} = 1$ TeV (middle line) and $\sqrt{s} = 3$ TeV (lower line). The parameter space above the lines corresponds to the observable events. Also plotted (solid lines): the experimental limits on U_{eN}^2 from $\beta\beta_{0\nu}$ and on $|U_{eN}|^2$ from invisible Z decays, and the perturbative unitarity bound. Here the parameter space above the lines are ruled out. (From [9]).

They find [10] that at LHC the mass range with significant sensitivity can be extended to $m_N = 10$ Gev – 400 GeV. The mixing parameter can be probed to 10^{-3} at a 2σ level way beyond the L3 95% C.L. bound [12], and reaching about 10^{-6} in the low mass region at a 5σ level. Their results for LHC are shown in Fig. 6.

5.4. Like-Sign Leptons in t and W Decay

Bar-Shalom et al. [13] consider rare lepton-number-violating top-quark and W-boson four-body decays to final states containing a same-charge lepton pair, of the same or of different flavors: $t \to bW^-\ell_i^+\ell_j^+$ and $W^+ \to JJ'\ell_i^+\ell_j^+$, where $i \neq j$ or $i = j$ and JJ' stands for two light jets originating from a $\bar{u}d$ or a $\bar{c}s$ pair. These $\Delta L = 2$ decays are forbidden in SM and may be mediated by exchanges of Majorana neutrinos. The corresponding diagrams are shown in Fig. 7.

They adopt a model independent approach for the Majorana neutrinos mixing pattern and calculate the branching ratios for these decays. They find that for $\mathcal{O}(1)$ mixings between heavy and light Majorana neutrinos (not likely but not ruled out) and if at least one of the heavy Majorana neutrinos has a mass of $\lesssim 100$ GeV, then the BR's for these decays are: $BR(t \to b\ell_i^+\ell_j^+W^-) \sim 10^{-4}$ and $BR(W^+ \to \ell_i^+\ell_j^+JJ') \sim 10^{-7}$ if $m_N \sim 100$ GeV and $BR(t \to b\ell_i^+\ell_j^+JJ') \sim BR(W^+ \to \ell_i^+\ell_j^+JJ') \sim 0.01$ if $m_N \lesssim 50$ GeV.

For such heavy Majorana neutrinos and assuming the dominance of only one heavy neutrino N the limits on its couplings to the charged leptons can be expressed in terms

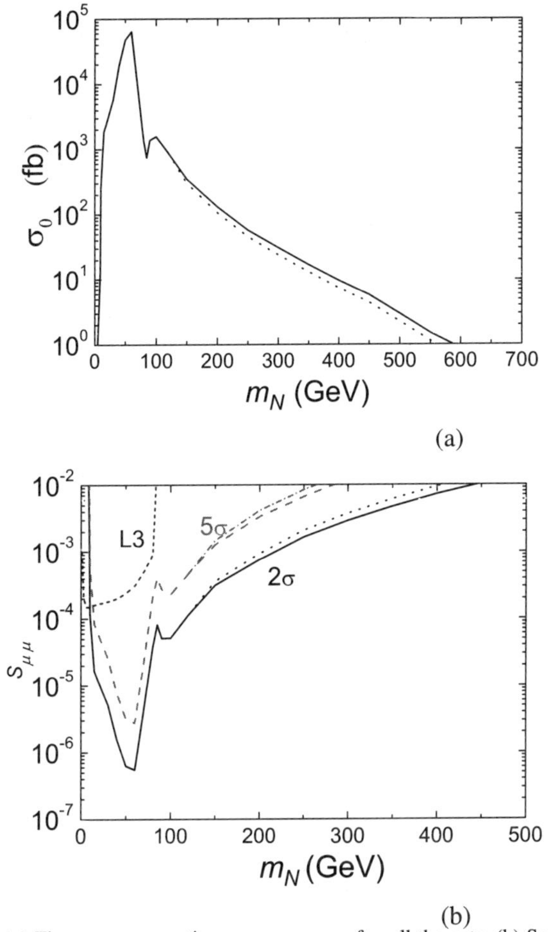

Figure 6. LHC results: (a) The raw cross section σ_0 versus m_N after all the cuts; (b) Sensitivity on $S_{\mu\mu}$ for 100 fb^{-1}. For comparison, the 95% bound from L3 search is included. The dotted lines tracing the curves illustrate the possible Higgs contribution of a mass 120 GeV. (From [10]).

of the products $\Omega_{\ell\ell'} \equiv B_{\ell N} B_{\ell' N}$, where $B_{\ell N}$ is the coupling of the W to a charged lepton ℓ and heavy neutrino N (see [14] and references therein). In particular, the limits on its flavor-diagonal couplings come from precision electroweak data, and at 90% CL are [15]:

$$\Omega_{ee} \leq 0.012 \; , \; \Omega_{\mu\mu} \leq 0.0096 \; , \; \Omega_{\tau\tau} \leq 0.016 \; , \tag{9}$$

while the limits on its flavor-changing couplings come from limits on rare flavor-violating lepton decays such as $\mu \to e\gamma$, μ, $\tau \to eee$ [14]:

$$|\Omega_{e\mu}| \leq 0.0001 \; , \; |\Omega_{e\tau}| \leq 0.02 \; , \; |\Omega_{\mu\tau}| \leq 0.02 \; . \tag{10}$$

Taking into account the present limits on the neutrino mixing parameters, they obtain more realistic values for these BR's: $BR(t \to b\ell_i^+ \ell_j^+ W^-) \sim 10^{-6}$ and $BR(W^+ \to$

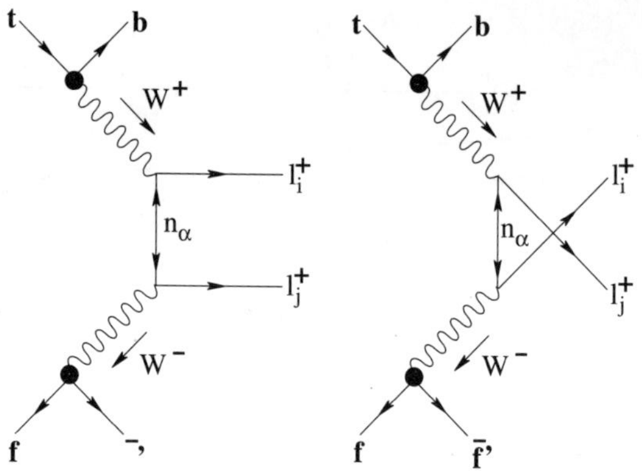

Figure 7. Feynman diagrams (t and u-channel) based on the kernel process $W^\pm W^\pm \to \ell_i^\pm \ell_j^\pm$, via Majorana neutrinos (n_α, $\alpha = 1-6$) exchanges. For $t \to b\ell_i^+\ell_j^+W^-$, cut the lower W-boson line and, for $W^+ \to \ell_i^+\ell_j^+ f\bar{f}'$, cut the upper W-boson line. (From [13]).

$\ell_i^+\ell_j^+ J\bar{J}') \sim 10^{-10}$ for $m_N \sim 100$ GeV and $BR(t \to b\ell_i^+\ell_j^+ J\bar{J}') \sim BR(W^+ \to \ell_i^+\ell_j^+ J\bar{J}') \sim 10^{-6}$ for $m_N \lesssim 50$ GeV.

Table 1. $BR(t \to bW^-\ell_i^+\ell_j^+)$ in the various $\ell_i\ell_j$ channels, subject to the bounds in (9) and (10) on the neutrino mixing parameters, for $m_N = 90$ and 100 GeV. For the flavor changing channels it is assumed that $\Omega_{\ell\ell'} = \sqrt{\Omega_{\ell\ell}} \times \sqrt{\Omega_{\ell'\ell'}}$ in case the limits on the couplings in (10) are weaker than the individual limits given in (9). (From [13]).

	$BR(t \to bW^-\ell_i^+\ell_j^+) \times 10^6$					
$\ell_i\ell_j =$	ee	$\mu\mu$	$\tau\tau$	$e\mu$	$e\tau$	$\mu\tau$
$m_N = 90$ GeV	1.4	1.1	1.9	$1.1 \cdot 10^{-4}$	1.6	1.4
$m_N = 100$ GeV	0.6	0.5	0.8	$0.4 \cdot 10^{-4}$	0.7	0.6

In the case of the top-quark decay $t \to bW^-\ell_i^+\ell_j^+$, taking $m_N \sim 100$ GeV and using the limits from (9) and (10), the BR's for the various $\ell_i^+\ell_j^+$ channels are given in Table 1.

6. Conclusions

$0\nu\beta\beta$ if seen would imply an interesting scenario in which lepton number is violated by two units. This LNV by two units could be because of a tree-level Majorana mass for one or more neutrinos, or because of some other mechanism, which in turn would also generate a Majorana neutrino mass. LNV by two units could be seen in other experiments. In particular, some possibilities at colliders suggested recently have been reviewed.

References

[1] There exist a number of reviews on this topic. See for example, H.V. Klapdor-Kleingrothaus and U. Sarkar, Proc. Ind. Nat. Sc. Acad., **70 Part A**, 251 (2004); S. Elliott and J. Engel, J. Phys. **G30**, R183 (2004); M. Hirsch, arXiv:hep-ph/0609146; P. Vogel, arXiv:hep-ph/0611243; U. Sarkar, these proceedings.

[2] J. Schechter and J. Valle, J. Schechter and J. W. F. Valle, Phys. Rev. D **25**, 2951 (1982).

[3] M. Hirsch, S. Kovalenko and I. Schmidt, Phys. Lett. B **642**, 106 (2006) [arXiv:hep-ph/0608207].

[4] R. N. Mohapatra, Phys. Rev. D **34**, 909 (1986); M. Hirsch, H. V. Klapdor-Kleingrothaus and O. Panella, Phys. Lett. B **374**, 7 (1996).

[5] M. Hirsch, H. V. Klapdor-Kleingrothaus and S. G. Kovalenko, Phys. Rev. D **53**, 1329 (1996); M. Hirsch, ref. [1].

[6] R. N. Mohapatra, Phys. Rev. D **34**, 3457 (1986); J. D. Vergados, Phys. Lett. B **184**, 55 (1987).

[7] M. Hirsch, H.V. Klapdor-Kleingrothaus and S.G. Kovalenko, Phys. Rev. **D54**, 4207 (1996).

[8] G. Rizzo, Phys. Lett. **B 116**, 23 (1982); G. Bélanger, F. Boudjema, D. London and H. Nadeau, Phys. Rev. **D 53**, 6292 (1996); G. Bélanger, *Proc "Lepton and Baryon number violation in Particle Physics, Astrophysics and Cosmology"*, eds. H.V. Klapdor-Kleingrothaus and I.V. Krivosheina, (Bristol, UK, Institute of physics), 605 (1998).

[9] D. Atwood, S. Bar-Shalom and A. Soni, pair arXiv:hep-ph/0701005.

[10] T. Han and B. Zhang, Phys. Rev. Lett. **97**, 171804 (2006).

[11] A. Datta, M. Guchait, and A. Pilaftsis, Phys. Rev. **D50**, 3195 (1994); F.M.L. Almeida, Y.A. Coutinho, J.A.M. Simoes and M.A.B. Vale, Phys. Rev. **D62**, 075004 (2000); O. Panella, M. Cannoni, C. Carimalo, and Y.N. Srivastava, Phys. Rev. **D65**, 035005 (2002).

[12] L3 Collaboratioin: O. Adriani et al., Phys. Lett. **B295**, 371 (1992); P. Achard et al., Phys. Lett. **B517**, 67 (2001).

[13] S. Bar-Shalom, N. G. Deshpande, G. Eilam, J. Jiang and A. Soni, Phys. Lett. **B643**, 342 (2006).

[14] S. Bray, J.S. Lee and A. Pilaftsis, Phys. Lett. **B628**, 250 (2005).

[15] S. Bergmann and A. Kagan, Nucl. Phys. **B538**, 368 (1999).

NEUTRINOLESS DOUBLE BETA DECAY
Editors: V.K.B. Kota and U. Sarkar

THE INDIA BASED NEUTRINO OBSERVATORY - CURRENT STATUS

V. M. Datar[1] (on behalf of the INO collaboration)

Nuclear Physics Division, Bhabha Atomic Research Centre, Mumbai - 400085, India

Abstract

The current status of the India based Neutrino Observatory (INO) is summarized. The two major physics goals are an unambiguous demonstration of neutrino oscillation including a more precise measurement of the associated neutrino parameters and a search for matter effects in neutrino oscillation, using the charge identification capability of the magnetized iron calorimeter, which would help determine the sign of one pair of neutrino mass differences. The status of the 1 m^3 prototype iron calorimetric detector, the design of the 50 kton magnet, the experience with resistive plate chambers used for tracking the charged particles produced in neutrino-iron interactions and the planned electronics and data acquisition system will be presented.

[1] **E-mail:datar@barc.gov.in**

1. Introduction

The neutrino was invented by Pauli [1] in 1930 to resolve the energy-momentum conservation and spin-statistics crisis in beta decay. The first evidence of the existence of the electron (anti)neutrino was provided in a pioneering reactor experiment [2]. This was followed by the discovery of the muon neutrino and, much later, the tau neutrino [3]. The helicity of the neutrino was shown [4] to be -1 $\pm$ 0.3 in agreement with the two component neutrino theory. An upper limit of the anti-neutrino mass was set at about 55 eV/c^2 through a careful measurement of the beta spectrum in tritium decay near the end point [5]. After unsuccessfully searching for neutrinos at a reactor (which is a copious source of *anti*neutrinos) [6] Davis used the radiochemical detection technique, involving the separation of ^{37}Ar from 600 tons of the cleaning fluid C_2Cl_4 containing ^{37}Cl, to measure neutrinos produced in nuclear reactions, and beta decays of the unstable nuclei produced thereby, in the hot core of the sun [7]. The roughly threefold shortage came to be known as the solar neutrino problem [8]. One of the explanations proposed to explain this shortfall was that the electron neutrinos produced in the solar interior change into another type (flavour) of neutrino, which is not measurable by the ^{37}Cl based detector. This chameleon like behaviour, known as neutrino oscillation, was first proposed by Pontecorvo [9]. The solar neutrino problem was resolved in a definitive manner by the Sudbury Neutrino Observatory experiment [10] using 1 kton of heavy water. Charged current interactions measured the electron neutrinos while neutral current events measured all neutrinos, irrespective of their type or flavour $(\nu_e, \nu_\mu, \nu_\tau)$. The shortfall in the ν_e flux was recovered in the flux of $\nu_\mu + \nu_\tau$.

Table 1. Best values of neutrino parameters.

Parameter	Exp. value (1σ)
Δ^2_{21}	$(7.9\pm0.4)\times10^{-5}$ eV^2
Δ^2_{23}	$\pm2.4\pm0.2\times10^{-3}$ eV^2
θ_{12}	$34.1^\circ\ ^{+1.6^\circ}_{-1.2^\circ}$
θ_{23}	$41.6^\circ\ ^{+5.7^\circ}_{-2.9^\circ}$
θ_{13}	$<8^\circ$

An equally intriguing problem resulted from the detailed measurements of atmospheric neutrinos. The IMB [11] and Kamiokande [12] collaborations found an anomalous ν_μ/ν_e ratio as a function of zenith angle. This ratio is expected to be close to 2, for high energy neutrinos, and the same for all directions in the absence of oscillations. If ν_μ oscillates into ν_τ the above ratio would be 2 for down going neutrinos but smaller than 2 for upgoing neutrinos. SuperKamiokande provided the first definitive results [13] which showed that neutrinos oscillate and therefore possess a small mass. It may be mentioned that atmospheric neutrinos were first detected at Kolar Gold Fields by an Indian team, just ahead of an experiment led by Reines in a South African mine [14].

The above mentioned, as well as a few more, key experiments have led to a dramatic change in our understanding of neutrinos and cannot be understood within the hitherto successful standard model of high energy physics. The widely accepted explanation of the experimental observations is that neutrinos switch identities, or oscillate into other flavours, as they propagate. This can occur if the flavour eigenstates (electron, muon and tau) are not simultaneously mass eigenstates (m_1, m_2, m_3). In general these two bases are connected by a unitary matrix. For 3 active flavours there are 7 independent parameters *viz.* the 3 mass parameters, 3 mixing angles θ_{12}, θ_{13}, θ_{23} and δ_{CP}. If the neutrino is its own (Majorana) antiparticle there are two additional phases which, however, would not be observable in ν-oscillation experiments. Table 1 lists the presently known experimental values [15]. Here Δm^2_{ij} or Δ^2_{ij}, for short, is defined as $m^2_i - m^2_j$.

There are several experiments planned in the near future which will improve the precision further and might even throw up unexpected results. Neutrinos are providing us the first clues to physics beyond the standard model and their study is expected to provide even more surprises. With a view of reentering this exciting area an initiative has been taken to set up an underground laboratory in India. One of the main experiments is aimed at making precision measurements on atmospheric neutrinos using a large iron calorimetric (ICAL) detector.

Two other important problems in neutrino physics should be mentioned. The first of these is that we do not yet know whether the neutrino is its own antiparticle (Majorana) or not (Dirac). A definitive measurement confirming or disproving the claim [16] of observing neutrinoless double beta decay (DBD) (which, if true, makes the neutrino a Majorana particle) would be crucial. Moreover, if the neutrino is a Majorana particle, the rate of 0ν DBD can be related to its absolute mass. The second important problem is to measure the absolute mass of the neutrino irrespective of whether it is a Majorana or Dirac particle. Presently the upper limit derived from a beta spectral measurement of tritium is 2.2 eV/c^2 [17] and there are plans to search for a mass down to about 0.2 eV/c^2 [18].

2. Major Physics Goals of ICAL

The major physics issues and questions that the ICAL detector will address are: (1) Observation of oscillatory pattern of muon neutrino flux with L/E and a precise measurement of neutrino oscillation parameters. (2) Search for matter effects leading to a determination of the sign of Δm^2_{23}. This depends crucially on the magnitude of θ_{13} which will be determined by other ongoing/planned experiments. (3) Whether θ_{23} is 45°; else determine whether $<$ or $>$45° (octant ambiguity). (4) Measure leptonic CP phase δ_{CP}, if θ_{13} is non-zero. Item no.2 can be studied using atmospheric neutrinos provided θ_{13} is $>$ about 5°. The sensitivity to the ordering of the neutrino masses arises as follows. The $\nu_\mu - \nu_\tau$ and $\nu_\tau - \nu_e$ mixing changes, through the $\nu_e - e$ charged current interaction, both the mass and the mixing angle from their vacuum values to those including matter effects. The sign of the contribution to each of these is opposite for neutrino and antineutrino for a given Δ_{32}. For the opposite sign of Δ_{32} these contributions change sign allowing a discrimination between the the normal ($\Delta m^2_{32} > 0$) and inverted ($\Delta m^2_{32} < 0$) hierarchies. At a later stage this and items 3 and 4 could be studied even better with an accelerator produced neutrino beam and neutrino factory, respectively.

3. Choice of Detector and Site

The ideal neutrino detector should have as low a threshold as possible possibly sub-MeV if it should be sensitive to geo, nuclear reactor, solar, supernovae and atmospheric neutrinos. Keeping in mind the physics reach of a potential detector, the technical capability of our R&D and that of our industry, feasibility of fabrication and assembly of subsystems of the detector and the time required to make it while being competitive internationally, a large magnetized iron calorimeter (ICAL) measuring atmospheric neutrinos seemed to be the best option. A low energy threshold ICAL detector using thin iron sheets as the target material for neutrinos would increase its dimensions while the use of very thick iron plates would increase that threshold. While the choice of 6 cm iron plates in ICAL results in an energy threshold of a few hundred MeV atmospheric and accelerator produced muon neutrinos can be studied fruitfully while keeping its size reasonable. A proposal [19] for a similar detector, called MONOLITH, was made by an Italian group but was not funded. The need for such a detector with its inherent charge discrimination capability, allowing it to distinguish between μ^+ and μ^- produced through charged current interaction of ν_μ and $\bar{\nu}_\mu$, respectively, was stressed in a recent APS study [20]. Such a detector is complementary to the water Cerenkov detectors such as SuperKamiokande and smaller magnetized iron calorimetric detectors measuring neutrinos produced at accelerators at relatively short baselines, such as the MINOS experiment [21].

A magnetized iron calorimeter is an attractive option if charge identification of the muons following a charged current interaction with the target nucleus is necessary. A magnetic field of about a tesla can be rather easily obtained, using a suitably configured DC current carrying coil for excitation, in soft iron or steel with low carbon content. It is available in large quantities and is reasonably cheap.

A reasonably precise measurement of the energy-momentum of the interacting atmospheric neutrino is required to measure L/E on an event by event basis. This can be done through by tracking the charged muon, produced via the charged current neutrino-nucleus interaction, by many layers of a position sensitive detector. The directional information, up or down going, can be obtained for muons which may or may not stop in ICAL through a fast time (sequence) measurement of the individual detectors. The total area required to be covered is about 10^5 m^2. Of the two possibilities *viz.* gas detectors operated in the avalanche or streamer mode and plastic scintillators with fibre readout, a choice of the former was made. Among the various options in gas detectors the resistive plate chamber (RPC) seems to be a good choice due to its good position and time resolution, ease of construction in large numbers, ruggedness, low cost/unit area, and operational experience in other large experiments. The R&D work on the glass RPC detector including the associated subsystems involving gas circulation, electronics and data acquisition is ongoing.

In consultation with the Geological Survey of India two possible sites for INO were identified, one at Rammam near Darjeeling in West Bengal and another at PUSHEP, Masinagudi, near the foothills of the Nilgiris in Tamilnadu. Two teams made an in depth study of the advantages and shortcomings of each site. A site selection committee consisting of physicists, a civil engineer and geologists was formed. Criteria were framed to decide the suitability in terms of depth, seismicity, proximity to industrial centres, access by road etc. This committee looked into all aspects and recommended, unanimously, that Masinagudi was the preferred site for INO.

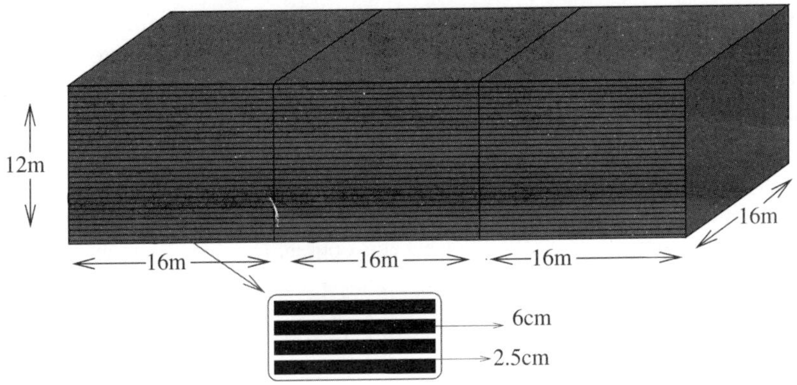

Figure 1. Schematic view of the 50 kton iron calorimeter detector consisting of 3 modules each having 140 layers of iron plates.

4. Iron Calorimeter and its Subsystems

The conceptual design of ICAL has been made and has a modular structure. Each module weighing 16 kton has a size of 16 m$\times$16 m$\times$12 m and comprises of 140 layers of a unit cell consisting of a 6 cm thick soft iron/low carbon steel plate and a 2.5 cm layer of a X-Y position sensitive resistive plate chamber (RPC). The magnetic field generated with the help of two sets of DC-current carrying coils will be $\sim$ 1 tesla. A schematic of the 50 kton detector is shown in Fig. 1.

Some more details of each of the subsystems including the present status are given in the following sections.

4.1. Magnet

The main design criteria for the ICAL magnet were piecewise field uniformity, modularity, optimum copper-to-steel ratio and access for maintenance. A preliminary design of the 16 kton module has been made including a finite element calculation of the magnetic field, using the commercial software Magnet 6.0 [22], for a few configurations of the current carrying coils. Based on these calculations the presently preferred configuration was arrived at. More details may be found in Ref. [23].

4.2. Resistive Plate Chamber

The resistive plate chamber consisting of two glass plates 2 m$\times$2 m area and 2 mm thick separated by an insulating spacer of 2 mm will be used. This choice of glass, as opposed to bakelite, was based on considerations such as cost, availability, ease of construction and suitability for an underground experiment involving low event rates. Most of the R&D work has been carried out using RPCs 30 cm$\times$30 cm in size but chambers of a larger size using locally made float glass (1.2 m×0.9 m) and Japanese float glass (1 m×1 m) have also been made. The gas mixing and circulation system has been developed in collaboration with a local vendor and, apart from a few teething problems, has performed quite satisfactorily. Similarly vendors for the conductive coating on the outer walls for applying the high voltage, polycarbonate spacers and buttons and gas inlet/outlet connections were identified and have delivered the items to specification in the smaller numbers required for the prototype detectors.

A test stack with 10 small sized RPCs (30 cm×30 cm) was set up at the RPC laboratory at TIFR and used to track cosmic muons. The trigger was provided by plastic strip scintillator detectors placed above and below the RPC stack. Fig. 2 shows a picture of the setup. The RPCs have been operated in the streamer mode, with

Figure 2. Muon tracking setup using RPC stack.

a gas mixture of HF134a:Ar:isobutane of 62:30:8, or in the avalanche mode with HF134a:isobutane of 95.5:4.5. The electronic pulse following a minimum ionizing particle traversing the RPC is a few hundred mV across a 50Ω load when operated in the streamer mode, while the corresponding figure for the avalanche mode is between 1-5 mV. While the streamer mode of operation simplifies the electronics, since the pulse can directly trigger a fast timing discriminator, the counters have not functioned in this mode for periods beyond about a month. The reason could be the formation of HF in the presence of moisture as an impurity in the detector gas. On the other hand two RPCs have been operating with individual efficiencies of 75-85% in the avalanche mode for a period of more than 18 months. Both these counters have dimensions of about 30 cm×40 cm and have been made from Japanese float glass. The problem of short lifetimes of RPCs [24] using local float glass could be because of some critical parameters and is being addressed through measurements of chemical composition, surface roughness as probed by reflectivity and atomic force microscopy etc. Other options such as using a freon-less detector gas are also being explored.

4.3. Gas Mixing and Distribution System

A 4-gas mixing and distribution unit suitable for the test bench as well as for the prototype 1 m^3 detector was designed and developed locally (see Fig. 3). It can supply gas at slightly above 1 bar to 16 detectors and can be operated in a continuous flow mode. The important features of this system are an input gas purifier (to remove oil and moisture traces) and 2 μm dust filters, a gas mixing system using mass flow controllers, flow sensors and monitors. In addition there are moisture monitors, safety bubblers on individual gas lines to prevent excess-pressure in the RPCs, isolation bubblers using low vapour pressure silicone oil preventing air, into which the gas is vented, from back diffusing into the RPC, an exhaust manifold and a remote control and monitoring system with a PC interface.

Figure 3. Front view of the 4-component gas mixing unit with 16 output channels.

The designing of a closed loop gas system for the 50 kton detector has been initiated. The system would be capable of mixing upto 4 gases, have a purification column to remove trace amounts of moisture and detector gas breakdown products, gas manifolds, flow controllers (mass flow controllers and impedances), sensors for measurement of various gas parameters such as flow, temperature, pressure, humidity etc and their logging in PC based readout and acquisition system.

4.4. Electronics and Data Acquisition System

The passage of a minimum ionizing particle induces a voltage pulse on the corresponding X- and Y- pickup strips. This pulse (with or without amplification, depending on whether the RPC is operated in the avalanche or streamer modes) goes to a fast timing discriminator (TD) located at the end of the strip. The logic output of the TD is then used to tag the strip which was 'hit' and also given to a multiplexed TDC for recording the time information. An FPGA based trigger module fabricated in-house recording the hit pattern and multiplexing will be used in the prototype detector tests. A multi-level programmable trigger generator with physics motivated trigger logic will be used to initiate data recording. The data acquisition system (DAQ) will be based on the VME standard and will be linked to PCs with the Linux OS. More details may be found in the Ref.[23]. The present plan is to use a scaled up version of this system but experience with the 1 m^3 prototype detector should provide inputs for the ICAL detector electronics and DAQ.

4.5. ICAL Prototype

A detector on the scale of ICAL has never been built in the country before. It is therefore prudent to build a smaller version and gain experience with the various subsystems and make course corrections on the way towards the design and fabrication of the 50 kton detector. The INO collaboration decided to build a $\sim$2.6 m$\times$2.8 m$\times$2.5 m ICAL prototype detector with a $\sim$1m^3 fiducial volume of detector. This detector will track cosmic ray muons.

5. Simulation of ICAL

The simulation of the 50 kton ICAL has been done in three steps. Firstly, the detector geometry was defined using the GEANT [25] detector simulation software from the CERN library. Neutrino events were generated using NUANCE [26]. Finally the simulated data was used to reconstruct particle tracks and their energy-momentum. The data can now be projected for analysis and visualization. A sample plot is shown in Fig. 4 which compares the allowed region in θ_{23}-Δm^2_{23} space from ICAL simulations with SK results [27].

6. Training People for INO - The First INO School

There is a need to develop expertise among the younger group of people who will be associated with the INO project for the next 10 years or more. This will also provide a strong base for future high energy projects both national and international. Various ideas are discussed in the INO Project report such as the formation of an INO School modelled on the lines of the successful BARC Training School, direct recruitment of

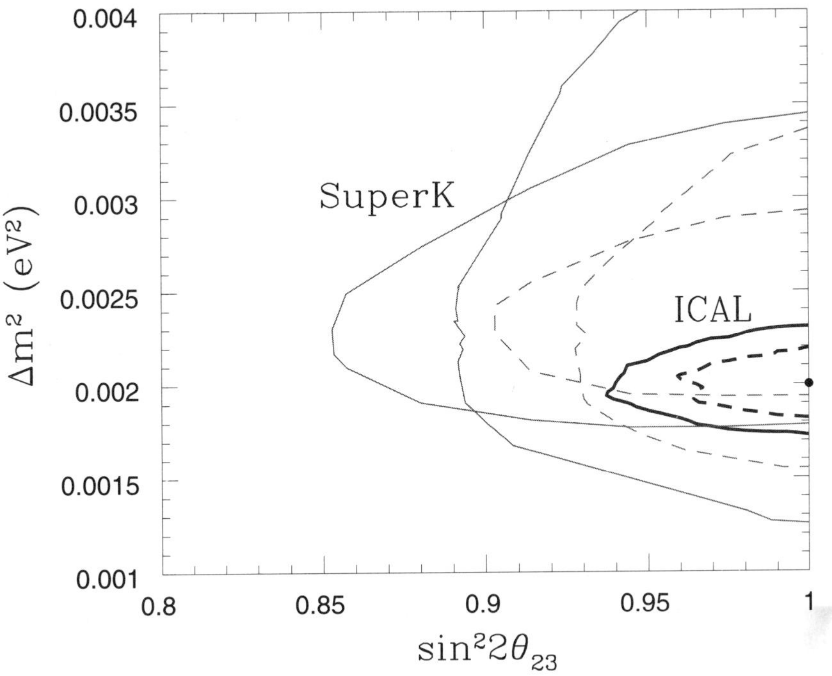

Figure 4. Allowed parameter space from a 300 kton.yr ICAL simulation using fully contained CC muon events (dashed-90% CL, full-99%CL) compared with SK results. The input parameters used in the simulation are indicated by the dot in the figure.

physicists and engineers and instituting INO Fellowships at selected universities, engineering colleges, NIITs and IITs. A modest beginning in the first mentioned approach has been made this year. A 4 week intensive program involving about a dozen students and recently recruited staff (with a physics background) doing INO related work at the various centres was completed in April-May of 2006. The theoretical component was organized and conducted by HRI, Allahabad while the experimental portion took place at VECC, Kolkata. The feedback was good and hence encouraging.

7. Estimated Cost and Time Schedule

The INO project has been put up through the Megasciences Committee of the Department of Atomic Energy for the 11th five-year plan. The INO project report [23] has most of the details as of May 2006. The total projected cost is Rs. 670 crores ($\sim$ \$150 M) over 2 five-year plan periods. About half of this estimated budget goes toward the low carbon steel for the 100 kton magnet.

The detailed project report (DPR) for the site and associated infrastructure, including that necessary for the underground and overground laboratories, is being prepared by the TNEB (Tamilnadu Electricity Board) engineering team. This contract was given by an Engineering Task Force set up by Chairman, DAE. This DPR is in the final stages of completion and should be ready by the end of May, 2007. An Engineering Design Report will also be prepared for ICAL.

An approximate timeline is that the first phase, involving the setting up of the 50 kton detector, should be completed 7 years from the time of financial sanction. By then other experiments such as double-CHOOZ would have measured or put an upper limit on θ_{13}. If this parameter, which is crucial to the existence of matter effects that would be measurable by ICAL, is not too small the next phase of adding additional modules totalling 50 kton would be undertaken. The construction time should be smaller for the second phase, perhaps ~ 3 years. Efforts are also being made to seek international participation in this experiment. If such collaborations materialize also with accelerator based groups the possibility of doing long baseline neutrino measurements, with much higher sensitivity to some of the neutrino parameters, might become a reality.

8. Current Status of INO

An interim report of the INO project was presented in May 2005 to the Chairman, DAE. Another presentation was made to the Scientific Advisory Committee to the Prime Minister in September, 2005. It was well received and a specific suggestion made to get it peer reviewed by international experts. The report was sent to seven distinguished scientists including six internationally well known experts in neutrino physics and one from India. The reports are generally very positive. Most referees have urged speedy construction in view of potential competition and another has expressed concern about the availability of adequate human resources. In any case, the collaboration is seeking a stronger participation, both national and international. The INO project is also one of the two mega-projects identified by the high energy physics community in the DAE-DST Roadmap meeting for high energy, nuclear and astrophysics held in April, 2006 at HBCSE, Mumbai. The INO collaboration looks forward to formal financial sanction by the Planning Commission and the Govt. of India by the 2nd half of 2007.

References

[1] W. Pauli, letter to participants of workshop at Tubingen (1930).

[2] C.L. Cowan and F. Reines, Science **124**, 103 (1956).

[3] G. Danby *et al.*, Phys. Rev. Lett. **9**, 36 (1962); K. Kodama *et al.* (DONUT collaboration), Phys. Lett. **B 504**, 218 (2001).

[4] M. Goldhaber, L. Grodzins and A.W. Sunyar, Phys. Rev. **109**, 1015 (1958).

[5] K.E. Bergkvist, Nucl. Phys. **B 39**, 317 (1972).

[6] R. Davis, Phys. Rev. **97**, 766 (1955); *ibid* Bull. Am. Phys. Soc., Washington meeting (1959).

[7] R. Davis, 2002 Physics Nobel prize lecture and references therein (http://nobelprize.org).

[8] J.N. Bahcall, arXiv:physics/0411190.

[9] B. Pontecorvo, Zh. Eksp. Teor. Fiz. **33**, 549 (1957); *ibid* **53**, 1717 (1967).

[10] Q. R. Ahmad *et al.*, Phys. Rev. Lett. **87**, 071301 (2001); *ibid* **89**, 011301 (2002); *ibid* **92**, 181301 (2004).

[11] T.J. Haines *et al.*, Phys. Rev. Lett. **57**, 1562 (1986); D. Casper *et al.*, Phys. Rev. Lett. **66**, 2561 (1991).

[12] K.S. Hirata *et al.*, Phys. Lett. **B 205**, 416 (1988); *ibid* **B 335**, 237 (1994).

[13] S. Fukuda *et al.*, Phys. Rev. Lett. **81**, 1562 (1998); *ibid.* **82**, 2644 (1999); *ibid.* **85**, 3999-4003 (2000).

[14] C.V. Achar *et al.*, Phys. Lett. **18**, 196 (1965); *ibid.* **19**, 78 (1965); F. Reines *et al.*, Phys. Rev. Lett. **15**, 429 (1965).

[15] G.L. Fogli, E. Lisi, A. Marrone, A. Palazzo and A.M. Rotunno, arXiv:hep-ph/0506307

[16] H.V. Klapdor-Kleingrothaus, I.V. Krivosheina, A. Dietz and O. Chkvorets, Phys. Lett. **B 586**, 198 (2004).

[17] V.M. Lobashev *et al.*, Nucl. Phys. **A 719**, 153 (2004); C. Kraus *et al.*, Eur. Phys. J. **C 40**, 447 (2005).

[18] KATRIN collaboration, A. Osipowicz *et al.*, hep-ex/0109033; L. Bornschein, Nucl. Phys. **A 752**, 14c (2005).

[19] N. Y. Agafonova *et al.*, *MONOLITH: A massive magnetized iron detector for neutrino oscillation studies,* LNGS-P26-2000; http://castore.mi.infn.it/~monolith/ .

[20] APS Multidivisional Neutrino Study - Joint Study on the Future of Neutrino Physics: The Neutrino Matrix (Nov., 2004).

[21] D.A. Petyt, seminar at Fermi Lab, 30 March 2006, http://www-numi.fnal.gov/talks/results06.html.

[22] Magnet 6.0 finite element electromagnet software from Infolytica, Canada.

[23] http://www.imsc.res.in/~ino/OpenReports/INOReport.pdf; copies of the INO Project Report 2006 INO/2006/01 available on request from Prof. N.K. Mondal (nkm@tifr.res.in).

[24] S.S. Bhide *et al.*, Nucl. Phys. **B 158**, 195 (2006).

[25] GEANT, CERN Program Library W5013 (http://wwwasd.web.cern.ch/wwwasd/geant/).

[26] The NUANCE Neutrino Generator, D. Casper, Nucl. Phys. Proc. Suppl. **112**, 161 (2002).

[27] Y. Ashie *et al.*, [Super-Kamiokande Collaboration], Phys. Rev. **D 71**, 112005 (2005).

References

NEUTRINOLESS DOUBLE BETA DECAY
Editors: V.K.B. Kota and U. Sarkar

FEASIBILITY STUDY OF NEUTRINOLESS DOUBLE BETA DECAY IN ^{124}SN

Vandana Nanal[1]

Tata Institute of Fundamental Research, Mumbai 400005, India

Abstract

Neutrino physics is one of the present day thrust areas in experimental as well as theoretical physics. The present experimental evidences regarding the neutrino mass strongly suggest physics beyond the standard model of particles and play an important role in astrophysics and cosmology. The mass and the nature of the neutrinos can be best extracted through the study of nuclear (single) beta decay and double beta decay (DBD). With the upcoming INO laboratory in India, we propose to study the feasibility of an underground neutrinoless double beta decay ($0\nu\beta\beta$) experiment. This talk will focus on feasibility of a Sn bolometric detector for the study of $0\nu\beta\beta$ in ^{124}Sn.

[1] **E-mail:nanal@tifr.res.in**

1. Introduction

The terrestrial and extraterrestrial searches for neutrino mass through the study of atmospheric, solar, reactor and supernova neutrino sources have finally succeeded in establishing that neutrinos have mass [1]. The atmospheric and solar neutrino oscillation data suggest two types of mass spectra, namely, the normal hierarchy ($m_1 < m_2 < m_3$) and the inverted hierarchy ($m_3 < m_1 < m_2$) mass spectrum. However, the actual mass of neutrino cannot be extracted from these data. Determining the absolute neutrino mass scale would represent a major advance in our understanding of particle physics. The nuclear (single) β decay and double beta decay (DBD) can provide the information on absolute effective mass of the neutrinos. In particular, neutrinoless double beta decay is perhaps the only experiment that will tell us about the true nature of neutrino: whether it is a Dirac or a Majorana particle. The $0\nu\beta\beta$ decay mode will tell us if the total lepton number, one of the fundamental conserved quantities of the Standard model, is violated. The mass and nature of neutrinos play an important role in theories beyond the standard model, namely GUTs, supersymmetry, supergravity and string theory. Further, the neutrino physics has important implications in stellar evolution, supernova explosion, galaxy formation, thermal history of the universe, big-bang nucleosynthesis, dark matter and dark energy. Hence, the study of $0\nu\beta\beta$ decay has important consequences in nuclear physics, particle physics, astrophysics as well as cosmology. Further, it is going to open up new research frontiers in theoretical physics as well as add new technological know-how to the experimental physics.

The implications of present studies about nuclear double beta decay, which is expected to proceed through two competing modes namely the $2\nu\beta\beta$ decay and $0\nu\beta\beta$ decay, are far reaching in nature. The half-life of $2\nu\beta\beta$ decay, which is a product of accurately known phase space factor and the appropriate nuclear transition matrix element (NTME) $M_{2\nu}$, has been already measured for about ten nuclei out of 35 possible candidates [2, 3]. Given the significance of the $0\nu\beta\beta$, there are nearly 30 planned and proposed experiments worldwide [4]. So far only the Heidelberg-Moscow experiment has reported a positive evidence for the $0\nu\beta\beta$ decay in ^{76}Ge [5]. However, this result is based on very low statistics and needs to be confirmed. More recently, the CUORICINO experiment has placed a lower limit on $0\nu\beta\beta$ in ^{130}Te [6]. The existing and planned proposals to search for $0\nu\beta\beta$ show a rich diversity of approaches with many novel techniques involving different areas of research. Majority of the experiments, based on direct counting methods, aim to measure the characteristic properties of $0\nu\beta\beta$ decay such as emission of two simultaneous electrons from the same vertex inside the source and/or the constancy of the sum energy of the two emitted electrons. There are two different types of direct $\beta\beta$ decay experiments namely (a) active source - in which the source and the detector are identical and (b) passive source - in which the source and detector are different. One of the main considerations for a $0\nu\beta\beta$ candidate is the high Q value, since the decay rate is proportional $\sim Q^5$ and the interference from natural radioactive background is less at higher energies. Moreover, as these events are expected to be very rare, both source and detector have to be large in size. Thus candidates having a large isotopic abundance are favourable.

In the last couple of years, efforts have been made to explore the possibility of $0\nu\beta\beta$ decay experiment in conjunction with the proposal to study the neutrino oscillations at INO. After reviewing various existing and proposed experiments worldwide, ^{124}Sn (Q = 2.28 MeV, 5.8% abundance) and ^{150}Nd (Q = 3.37 MeV, 5.6% abundance) were identified as possible choices. A crucial criterion for detector design is high energy resolution for a precision measurement of the sum energy of two electrons emitted in $0\nu\beta\beta$ decay. Of various existing and proposed experiments, the CUORE experiment (^{130}Te) [7], the only low temperature bolometric detector so far, and the Ge experiment [8] have the desired energy resolution (0.5% and 0.2%, respectively). Of the two candidates (namely, ^{124}Sn and ^{150}Nd), at present, only Sn can be made into a bolometric detector. Tin becomes superconducting below 3.7 K and is nonmagnetic. Thus at temperatures below 100 mK the specific heat of Sn is very low, having only lattice contributions, and can be used as a bolometric detector [9]. Whereas, since Nd has 4f electrons and hence a magnetic moment, in most of its compounds the specific heat is much higher at low temperatures. This makes the Nd compounds looked at hitherto unsuitable for bolometry. Keeping in mind the necessity of high energy resolution, the community decided to focus on the Feasibility Study of Neutrinoless Double Beta Decay in ^{124}Sn.

2. Sn Bolometer

A bolometer consists of a thermally isolated detection mass connected to a heat bath only through a *weak* heat link. When the particle to be detected is stopped within the detector, the energy of the particle is converted to thermal energy and with a short rise time, the entire detector reaches a higher temperature. The weak heat link then conducts the heat away to the heat bath, bringing the temperature back to the base temperature with a relatively longer relaxation time. Under the assumption that the rise time is much smaller than the decay time, the temperature rise of the detector can be computed by equating the energy of the particle to the change of the enthalpy of the detector between the base temperature and the maximum temperature of the pulse. Thus the primary design considerations for the bolometer are the heat capacity of the detector which directly determines its sensitivity, and the rise time/decay time that determine how accurately the energy information can be extracted from the heat pulses. If the thermalization time within the detector is much smaller than the time taken for the thermal sensor to equilibrate with the detector, then the rise time is decided by the contact resistance between the detector and the heat sensor. Whereas the decay time is solely determined by the weak heat link (which is in fact the lead wires of the sensor) between the sensor and the heat bath. For good energy resolution, it is essential that the rise time is restricted to much less than the decay time.

We have initiated an effort to make a small cryogenic bolometer using natural Sn. The work plan is to measure thermal pulses arising from fission fragments from a ^{252}Cf source incident on a $\sim$100 mg Sn sample kept at 100 mK. As the fission fragments have large energy ($\sim$100 MeV), the expected large temperature rise $\sim$45 mK would facilitate setting up of detector and readout. Figure 1 shows the simulated pulses for the ^{252}Cf source, collimated to give $\sim$1 fission/sec on the sample, for a decay time of 10 ms. The simulation also took into account the much more numerous alpha particles (6.2 MeV) emitted by the source, which can be used to determine the energy resolution of the detector. The sensor for the temperature measurement, its mounting on the Sn

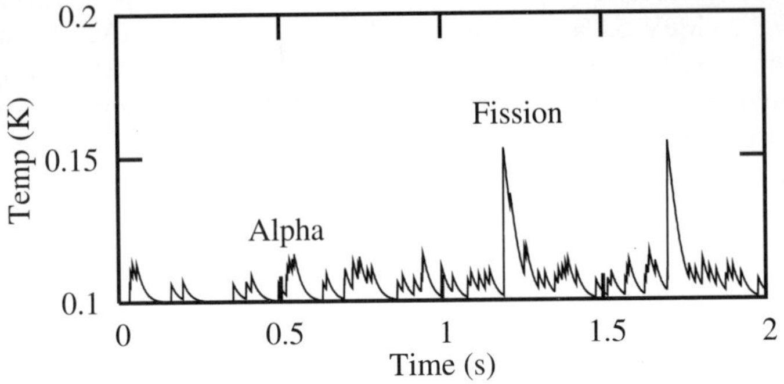

Figure 1. Simulated pulses on a 100mg Sn sample with a ^{252}Cf source (~1 fission/sec) assuming t_{decay}=10 ms and t_{rise}=0 ms.

sample and the design of thermal links for the readout at sub-Kelvin temperatures are some of the critical components of the detector. The experiment was carried out in a ^{3}He$-^4$He dilution refrigerator [10] at a base temperature of 100 mK. Since the available RuO_2 sensors had a relatively slow response time ~ms, it was necessary to design thermal links with a large time constant (~100 ms) to observe the thermal pulses. The choice of the link material thus becomes crucial as the total heat capacity of these links should be negligible compared to the Sn sample itself. One of the main difficulties is teh fact that all parameters like resistances, specific heat, thermal conductivity cahnge by several orders of magnitude in the tempearture range 20-200 mK. We have tried enameled copper wire (~50 μm dia), manganin wire (~150 μm dia) and found them to be unsuitable. Figure 2 shows a comparison of heat capacity at very low temperatures for 50 μm diameter copper wires of different lengths and Sn samples. Several trials are needed to optimize the design of links both in terms of the choice of material and lengths. The superconducting NbTi wire seems to be the most promising and will be attempted in later trials. Another crucial component, the sensor will be discussed in the next section.

Signal Processing Generally, the temperature change is deduced from the measurement of the resistance of the sensor.We have used, a commercially available bare RuO_2 thick film thermistor (RX102A-BR by LakeShore). It has a negative temperature coefficient with a sensitivity of about $0.2K\Omega/$mK at 100 mK. Its dimensions are about 1mm x 1mm x 0.4mm, and weighs 0.3 mg. Leads of enameled copper wire of diameter 40 μ m were carefully soldered onto the metal contact pads of the thermistor. In order to measure voltage drop across the heat sensor, a constant current is passed through it. It is important to note that such low temperature measurements are done with very low currents (~nA) to ensure that the associated power dissipation does not significantly affect the thermal load(~ 1 pW). The resistance of the RuO_2 sensor at a typical temperature of 200 mK is ~ 10 KΩ, which corresponds to a voltage ~ 100 μV. At this temperature, the sensitivity of the sensor ~100 Ω/mK. To achieve an accuracy of 1 mK in the temperature measurement, the error in the resistance measurement has

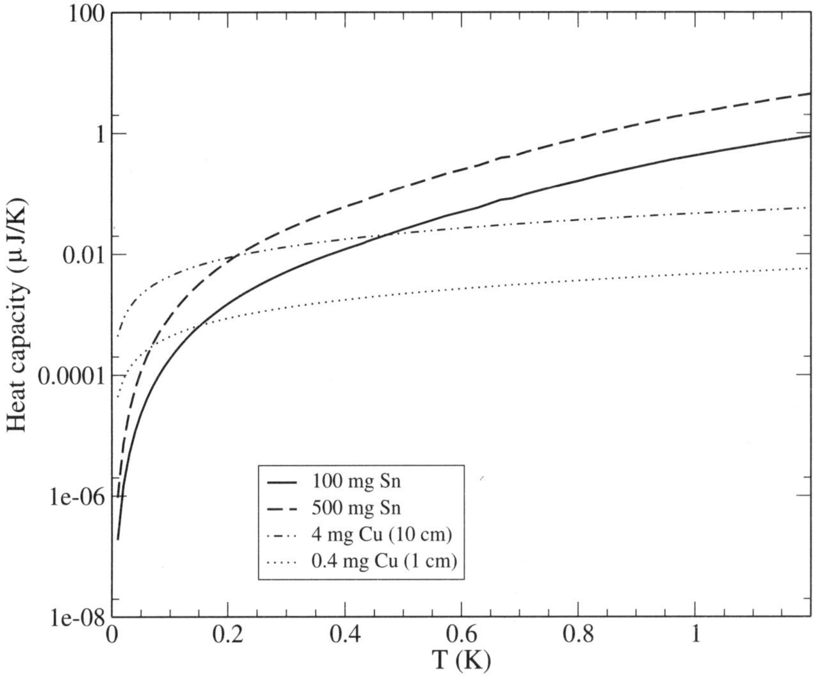

Figure 2. A comparison of Heat capacity of Sn and thermal links of 50μm diameter Cu of varying lengths [11, 12].

to be better than 100Ω, which in turn corresponds to an accuracy of 1 μ V in voltage measurement. This puts a very stringent limit on the the noise level in the setup, particularly since the signals need to be transmitted through several meters long cables as the readout electronics cannot be anchored to the cold station inside the refrigerator. Thus, it is essential to have suitable shielding for signal cables.Due to slow signal bandwidth, noise reduction at low frequencies is crucial. Another important aspect is the reduction of micro-phonics in the system, which requires whole setup to be kept isolated using anti-vibration platform. The measured voltage needs to be further amplified to $\sim$ 1 V level prior to digitization by an ADC. The readout electronics consisting of a high-gain differential pre-amplifier (gain $\sim$10000) with a matched pair of ICs was developed and successfully tested. The pulses were recorded using a CAMAC based data acquisition system (LAMPS). The measured heat capacity of the Sn sample is shown in Figure 3 using the above setup, where the lambda transition of the specific heat at its critical temperature is clearly visible. The optimization of the heat links and sensors with faster response are needed for recording thermal pulses and efforts are underway. Moreover, thermal equillibration times for a single crystal or polycrystalline sample will be significantly different and can be important for detector performance. This aspect also needs to be investigated.

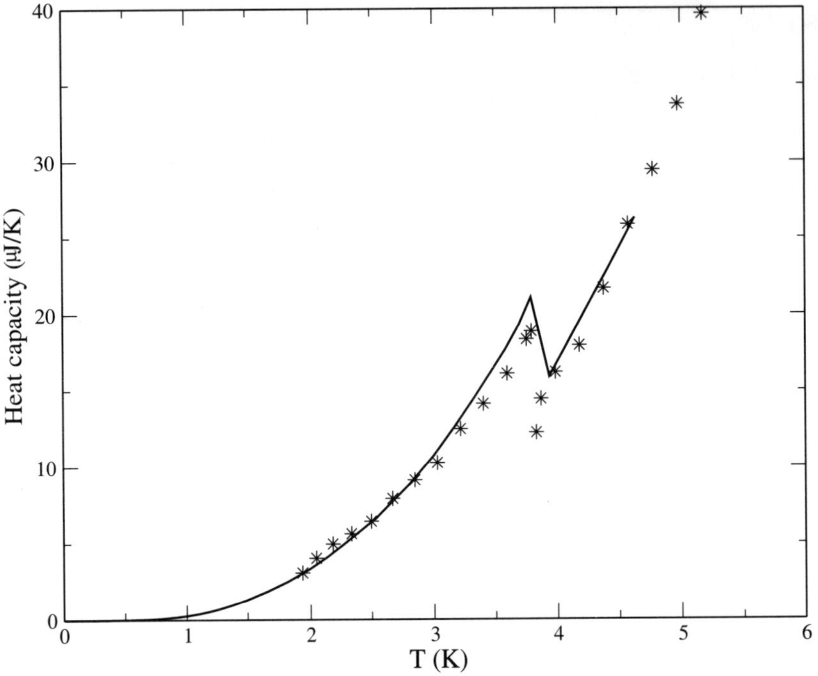

Figure 3. Measured heat capacity of Sn sample (*) showing the superconducting transition at 3.7 K together with extrapolated data from literature (solid line) [11].

3. Low Temperature Thermometry

As mentioned in the previous section, when energy deposited by the particle is converted into heat, a sensitive thermometer to record the temperature rise ΔT ($\sim \mu$K to mK) is necessary. The most generally used technique is to measure the change in resistance ΔR (either by ΔV or ΔI) and deduce the corresponding ΔT. An efficient bolometer requires large sensitivity (dR/dT) and low heat capacity. There are different types of sensors: Neutron Transmutation Doped (NTD) thermistors [13], Superconducting transition thermometer- Transition Edge Sensor (TES) and Superconducting Tunnel Junction (STJ) etc. Most of these are custom made and also would be required in a large quantity for a typical DBD experiment.

The NTD Ge sensors are widely used for high resolution mK measurements (e.g. CUORICINO [7]). The neutron transmutation doping is more homogeneous than chemically doped counterparts. The NTD Ge sensors are prepared by irradiation with thermal neutrons from reactors (typical flux $\sim$ 3.6X10^{18} neutrons/cm^2). The samples then are annealed at 400° C for 6 hours to repair the damage. Electrical contacts are later made on either side by implanting boron ions (3x10^{14}/cm^2 at $\sim$200 nm depth). A layer of palladium (20 nm) and gold (400 nm) is sputtered on surface and sample is further annealed for one hour at 250°C. The ^{70}Ge(21.2%) and ^{74}Ge(35.9%) undergo $\sigma(n,\gamma)$ reaction generating ^{71}Ge and ^{75}Ge, which subsequently decay to ^{71}Ga and

^{75}As, producing p-type Ge. The variation of resistance with temperature is given by

$$R(T) = R_0 exp(T_0/T)^{0.5} \tag{1}$$

where T_0 is related to the doping level and is found to increase with decreasing doping and R_0 is the intrinsic property of the parent Ge crystal.The NTDs also have fast response time. It is possible to produce NTD sensors at Dhruva (neutron flux ~2 x 10^{14} neutrons/cm^2/sec) or at Cirus (neutron flux ~4 x 10^{13} neutrons/cm^2/sec).

The Transition Edge Sensors (TES) are single layer or multilayer films operated near the transition temperature T_c [14]. As the temperature changes, film resistance will change drastically due to transition from superconducting to normal state. The advantage is that operating temperature can be suitably tailored as per the need. Many combinations like Mo+Au, Mo+Cu, Al+Cu, Ti+Au etc. are possible. However, these have a limited dynamic range and exhibit a nonlinear behaviour. Also the transition width affects energy resolution [15]. It is possible to fabricate TES by vacuum deposition of thin films at BARC or TIFR and work has been initiated for fabrication of molybdenium films.

4. Background Measurement

In rare event experiments like double beta decay ($T_{1/2} \sim 10^{20} - 10^{25}$ years), it is of paramount importance to identify and suppress unwanted background events, which limit the sensitivity of measurement. For observation time $t << T_{1/2}$, we can write

$$T_{1/2} \sim ln2(N_a t\varepsilon / N_{0\nu\beta\beta}) \tag{2}$$

where $N_a = N_A(M/A)i$ is number of $\beta\beta$ candidate atoms (N_A is Avogadro's number and i is isotopic abundance), ε is the detction efficiency. If the background rate is B (cts/kev/time), then background events in the window of interest can be written as

$$N_{bkg} = B\Delta Et \tag{3}$$

The sensitivity to $T_{1/2}$ can be estimated using $N_{0\nu\beta\beta} \sim \sqrt{N}_{bkg}$ as

$$T_{1/2} \sim ln2(N_A Mit\varepsilon / A(B\Delta Et)^{1/2}) \tag{4}$$

The whole class of background producing sources relevant to such experiments can be divided broadly into two categories. The first one is cosmogenical, which are produced by the interaction of cosmic rays with the atmospheric molecules and the materials available around the experimental setup. This is vastly reduced in underground laboratories. The second class of sources, which are most crucial for sensitivity, are the decay of radioactive nuclei present in the materials of detectors and its surroundings. Therefore, both the detector and shielding materials should have minimum radioactive contamination.

We have carried out a study of radiative background in the vicinity of 2.28 MeV (i.e. Q value for DBD in ^{124}Sn) from ambient background and various materials to be used in the bolometer using a 180 cc HpGe detector with 10 cm thick lead shield [16]. Most of the gamma rays in the ambient background originate from Uranium/Thorium decay chains ($T_{1/2} < 10^9$years). The characteristic γ-rays from various nuclei in these decay chains like ^{228}Th, ^{234}Th, ^{228}Ac, ^{235}U, ^{226}Ra, ^{214}Pb, ^{208}Tl, ^{214}Bi were clearly identified. The background observed in the region of interest (2268-2308 keV) was 3263$\pm$126 cts/keV/yr.

Recently, we have also carried out the background measurement at PUSHEP tunnel ($\sim$500 m depth) using a portable HPGe detector. Figure 4 shows a comparison of ambient background inside the tunnel and overground. The yield of most of the $\gamma-$rays in Uranium/Thorium decay chain show significant reduction inside the tunnel, whereas the primordial radioactivity, ^{40}K was found to be higher. In addition, the cosmogenic background was also measured inside the tunnel using a 2" x 2" NaI detector and was found to be lower by a factor of $\sim$ 40 in the region of 3 - 5 MeV.

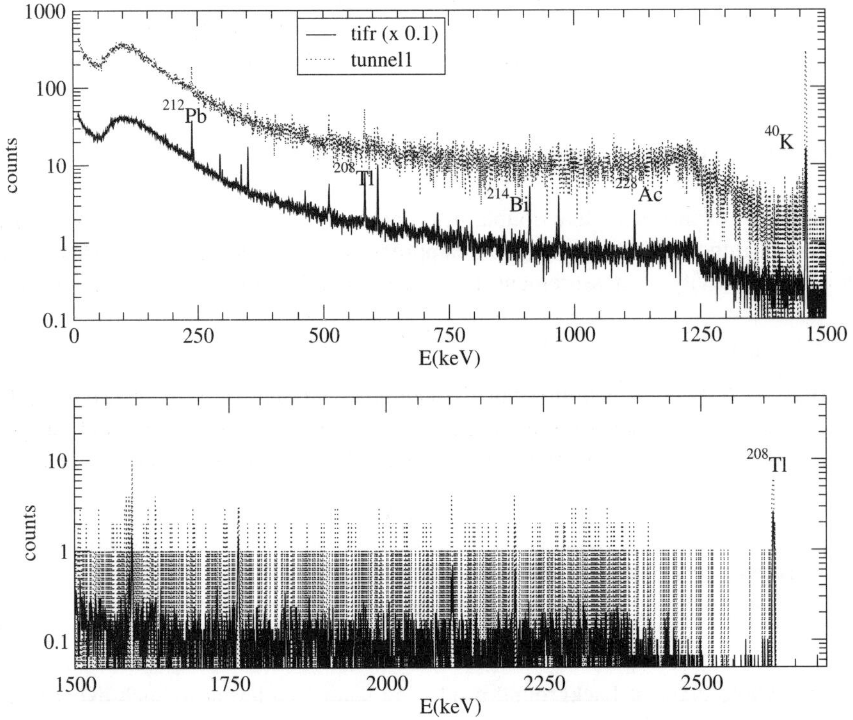

Figure 4. A comparison of ambient background at PUSHEP tunnel and overground (TIFR).

One of the problems for tin is that the Compton continuum of 2614.5 keV γ-ray, originating from ^{208}Tl decay, contributes in the region of Q value (2.288 MeV). Thus the suppression of 2614.5keV γ-ray is of paramount importance. To achieve the background level similar to *Cuoricino*, namely, 0.2 cts/keV/kg/yr, careful selection of low activity materials will be necessary and a combination of active and passive shielding will have to be designed. For a complete understanding of the background, extensive simulation studies will also be necessary.

5. Enrichment of Sn Isotope

As mentioned earlier, isotopic abundnace of ^{124}Sn is only 5.79%. Thus, it is essential to have enriched material or size of detector gets larger by order of magnitude. A preliminary survey of literature for Sn enrichment was carried out by Dr. Gantayet (L&PTD, BARC). The melting point of Sn is rather low, namely, 232°C and boiling point is 2602°C. Various methods employed for isotopic separation of tin are: Redox ion-exchange chromatography, Centrifugal extraction, electromagnetic separation and Laser based isotope separation (LIS).In particular, keeping in mind the need of large quantity of material, different LIS approaches were further investigated. The LIS is based on isotope selective excitation experiments. From the available spectroscopic data of ^{124}Sn, it is seen that single colour two photon ionization, two colour three photon ionization, single colour three photon ionization and three colour three photon ionization are some of the possible schemes that can be explored with existing Laser facilities at BARC. The proposed plan is to preapre a reactant feed as Van der-Waal complex of tin vapour and reactant molecule, selectively excite ^{124}Sn and remove enriched product by condensation.

6. Outlook: $0\nu\beta\beta$ Decay in ^{124}Sn

Initially, we propose to make a prototype of natural Sn of mass 0.5-1 kg. This would be set up at TIFR using a refurbished ^{3}He$-^4$He dilution refrigerator with a cooling power of 20 μW at 30 mK.This will also serve as a test bench for qualifying mK thermometry. From our present measurement of background as described in the preceding section, we expect to achieve a sensitivity of 6×10^{20} yrs with 1 kg natural Sn detector for an observation time of 1 year. The present limit in literature for $T_{1/2}$ for ^{124}Sn is 2.4 x 10^{18} years [17]. The extensive theoretical support for the calculation of nuclear transition matrix element (NTME), which is required for the extraction of neutrino mass (m_ν), exists within India and work is already in progress.

It can be seen from eqn. 4, that sensitivity improves linearly with mass of the detector. Based on preliminary NTME calculation of Rath [18] of $F_n \sim 2.754 \times 10^{-14}$yr^{-1} for ^{124}Sn and assuming 0.5% energy resolution with a background of $\sim$ 0.2 cts/kev/yr, similar to *Cuoricino*, we estimate that a large detector of $\sim$1 ton will be needed to be housed at INO underground laboratory to achieve a sensitivity of $m_\nu \sim$ 200 meV in 1 year observation time. The detector will be subdivided into several segments, size and number of which will have to be otpimized after detailed simulation studies. Suppression of cosmogenic and ambient background at underground laboratory, using both active and passive shielding will have to be taken into consideration. Novel techniques for discrimination of charged particles and photons will have to employed. Thus, we envisage a lab space of 5m x 5m x 7m for DBD studies in the main tunnel. Additional space for utilities (liq. He compressor, LN_2 storage, gas storage etc.) in the vicinity would be necessary. In addition, to minimize the contamination due to n-induced reaction, an additional underground laboratory for material processing and storage will be highly desirable.

Acknowledgement

This work is result of collaborative efforts with Prof. R.G. Pillay, Dr. V.M. Datar, Dr. A. Srivastava. Mr.Ravishankar Sundararaman (IIT, Kanpur), Mr. Vivek Singh (IIT, Kharagpur) and Mr. Himanshu Sharma (IIT, Kharagpur) have contributed to this work as a part of the summer projects. We are greatful to Dr.L.M. Gantayet (L&PTD, BARC) for helping with R & D of Sn enrichment, to Mr. J.N. Karande for help with electronics and to Prof. P.L. Paulose for use of the dilution refrigerator during initial tests. We thank late Prof. C.V.K. Baba for encouragement and support in initiating this project.

References

[1] INO interim report, 2005 and references therein. Also available at http://www.imsc.res.in/ ino.

[2] S. R. Elliott, A. A. Hann, M. K. Moe, Phys. Rev. Lett. **59**, 2020 (1987); ibid **59**, 1649 (1987); Nucl. Instrum. Methods A **273**, 54 (1988).

[3] S.R. Elliot and P. Vogel, Ann. Rev. of Nucl. and Part. Sci. **52**, 115 (2002).

[4] Report of the working group of the APS multidivisional neutrino study, arXiv:hep-ph/0412300, Dec.2004.

[5] H. V. Klapdor-Kleingrothaus, I. V. Krivosheina, A. Dietz, O. Chkvorets, Phys. Lett.B **586**, 198 (2004).

[6] C. Arnobaldi et al. Phys. Rev. Lett. **95**, 142501 (2005).

[7] C. Arnobaldi et al., Nucl. Instrum. and Meth. A **518**, 775 (2004).

[8] H.V. Klapdor-Kleingrothaus et al., Nucl. Instrum. and Meth. A **511**, 341 (2003), H. V. Klapdor-Kleingrothaus et al., Nucl. Instr. Meth. A **522**, 371 (2004).

[9] G. Forster et al. Journal of Low Temperature Physics, **93**, 219 (1993).

[10] Kelvinox, Oxford Instruments.

[11] C. A. Bryant and P. H. Keesom, Phys. Rev. **123**, 491 (1961).

[12] H. M. Rosenberg, Philosophical Transactions of the Royal Society of London. Series A, Mathematical and Physical Sciences, Vol. **247**, No. 933, 441-497 (1955).

[13] Eric Aubourg et al., Journal of Low Temperature Physics, **93**, 289 (1993).

[14] M. Frank et al., Journal of Low Temperature Physics, **93**, 213 (1993).

[15] I.J. Maasilta, Nucl. Instrum. and Meth. A **559**, 706 (2006)

[16] V. Singh et al., DAE-BRNS Symposium in Nuclear Physics, **51**, 328 (2006).

[17] Eric B. Norman and Dawn M. Meekhof, Phys. Lett. B **195**,126 (1987).

[18] P.K. Rath, Private Communication.

NEUTRINOLESS DOUBLE BETA DECAY
Editors: V.K.B. Kota and U. Sarkar

SIMULATION STUDIES ON BACKGROUND FOR DOUBLE BETA DECAY

A.K. Singh[1], S.K. Ghorui and P.K. Raina

Dept. of Physics & Meteorology, Indian Institute of Technology Kharagpur, Kharagpur 721302, India

Abstract

Monte-Carlo simulations are a widely used tool to help in designing new experiments and to study different aspects of it in nuclear and particle physics. In India effort are being made to plan an experiment to observe double beta decay process in ^{124}Sn using bolometric technique. We have started a program to simulate decay events produced by double beta decay process of ^{124}Sn as well as background events from radioactive decay processes in detector materials and surroundings. The event generator DECAY0 has been modified to generate decay events for various modes of the double beta decay of ^{124}Sn

[1] **E-mail:singh@phy.iitkgp.ernet.in**

1. Introduction

With the discovery of the fact that neutrino has finite rest mass by neutrino oscillation experiments like Super-Kamiokande [1] and SNO [2] and the recent one by Kamland experiment [3] efforts to measure mass of neutrino have increased world wide by various groups. Neutrinoless double beta decay is a unique phenomenon which not only can establish mass of neutrino quantitatively but also confirm the fact that neutrinos are Majorana particle. Several experiments are running on at different labs using different techniques and detectors systems, however, so far no success has been achieved. Due to large uncertainty in results only limits have been put for the mass of neutrino and best value is from the Heidelberg-Moscow collaboration [4, 5]. The difficulties related to experimental observation of the decay processes arise due to longer half life of the process which is $\sim 10^{25}$ years. This requires large mass of the decaying nuclei and longer time of measurements. Furthermore, half life of radioactive nuclei present is nature is of the order of 10^9 years which is much shorter compared to that of the the double beta decay (DBD) process. Therefore, any experiment related to double beta decay measurement should have an environment where background radiation due to other events are as small as possible. It is preferred to have labs under the mountain or in the mine in order to reduce background radiation produced due to Cosmic rays. Also, the materials which are used to make detectors as well as any component of the experimental setup and surroundings must be from radiopure materials. Therefore, it becomes a very important task to detect and finding out of the sources of background radiations for any DBD experiments. Simulation of backgrounds helps to identify and to suppress unwanted background events in order to improve the sensitivity. Thus it is essential to have an understanding of various issues related to the background prior to the designing a double beta decay experiment.

2. Sources of Background Radiations

The whole class of background producing sources relevant to such experiments can be divided broadly into two categories. The first one is cosmogenical, which are produced by the interaction of cosmic rays with the atmospheric molecules and the materials available around the experimental setup. When the detector and other materials are exposed to cosmic rays during production, storage and transport short lived radio isotopes are produced inside the material. This is vastly reduced in underground laboratories and one should wait for sufficient longer time to decay out these cosmogenic isotopes. The second class of sources, which are most crucial for sensitivity, are the decay of radioactive nuclei present in the materials of detectors and its surroundings. In this category important components consist of primordial activities of ^{238}U, ^{232}Th and from ^{40}K and also from anthropogenic radionuclides ^{137}Cs, ^{134}Cs, ^{125}Sb and ^{207}Bi released in nature during nuclear weapon tests and reactor accidents. Apart from discrete transition from these nuclei there are also radiations which produces continuous background in the energy spectrum. Important contributers in this category are Bremsstrahlung of ^{210}Bi, elastic and inelastic neutron reactions and radiations induced by direct muon hits. In the underground labs radioactive Radon gas is available resulting from decay chains of U and Th present in the rocks. The experimental setup therefore must be flushed with radio pure nitrogen or any other radio pure inert gas.

3. Simulation of Background

3.1. GEANT4 - a Simulation Toolkit

Since it is nearly impossible to eliminate all the sources of background, a detailed simulation of the background in the detector becomes an important part of such an experiment. The goal of simulation program is to build up a model for the background and to study the response functions of the detectors used in the measurements. In response to this, a new object-oriented simulation toolkit, Geant4 [6], has been developed. The toolkit provides a versatile and comprehensive software package for modern simulation applications that involve interaction and passage of particles through matter. It can handle setups with simple geometries to full-scale detector simulations for experiments at the Large Hadron Collider and other facilities efficiently and compactly, and allows visualization of the geometry and particle tracks through a variety of interfaces. It provides simulation for a wide range of physics processes based on theory, data or parameterisation. For example, hadronic interactions from thermal energies up to 1 PeV; electromagnetic interactions of charged hadrons, ions, leptons and photons from 250 eV to 1 PeV or more, as well as the production and propagation of optical photons. The implementation of the toolkit in an object-oriented design allows it to be easily extended, where appropriate, to meet the requirements of the user, through class inheritance. In addition, there are a many utilities available for visualization and analysis of resulting data. The graphics output may be either displayed immediately in a window, or written to a file in order to be viewed by external programs.

GEANT4 has modular and hierarchical structure where subdomains are linked by a uni-directional flow of dependencies. The key domains of the simulation of the passage of particles through matter are: geometry and materials; particle interaction in matter; tracking management; digitisation and hit management; event and track management; visualisation and visualisation framework; user interface.

To start a simulation of any decay process an event generator is required which describes initial kinematics of events, namely, types of particles, the decay processes, their energies, direction of movements and time of emission. In the following we descibe in brief about the event generator which we have used to generate events for the simulation of background raditions as well as for the DBD process of ^{124}Sn.

3.2. Event Generator

We have used an event generator "DECAY0", a Monte Carlo generator for the simulation of double beta processes and decays of radioactive nuclei provided by V. Tretyak [7]. It is also used in many other double beta experiments elsewhere. The working priciple of the event generator can be divided into two sections. In the first part it reads all the parameters of the nucleus and its decay needed for simulation from a database. The second part consist of a Monte Carlo events generator. The relevant data required to generate the decay process have been obtained from the database ENSDF, NUDAT, EADL [8] which includes decay modes, their probabilities and energy release, isotopes half-lives, radiation type, particle energy and intensity, parameters of nuclear levels and of nuclear transitions. It also uses data on atomic properties of isotopes, e.g. electron binding energies, electron capture subshell ratio, X rays and Auger electrons

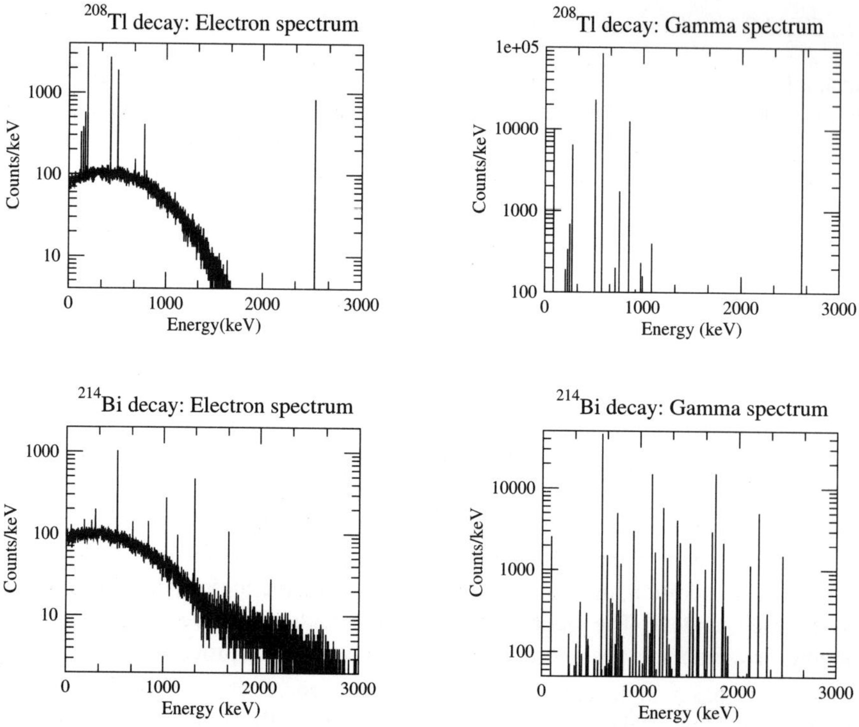

Figure 1. Histogram plot of electron and gamma decay of ^{208}Tl and ^{214}Bi isotopes. The events were generated using the event generator DECAY0.

intensities. The event generator part of the DECAY0 generates energy, direction and time of emission of the particles like, electrons and positrons for single and double beta decay; α, proton and neutron decays, γ from nuclear deexcitation process; conversion electrons; e^-e^+ pairs from internal pair conversion; γ quanta from bremsstrahlung in β decay and EC; neutrinos (antineutrinos) from EC or β (2β) decay; X rays and Auger electrons from the atomic deexcitation process. A typical spectrum obtained from the event generator for the decay processes of radioactive nculei ^{214}Bi and ^{208}Tl is shown in the Fig. 1.

The Decay0 can also generate events related to various modes of double beta decay processes of more than twenty nuclei. The program starts with a an interactive menu which can generate events of double beta decay process and has an option to generate events of radioactive decays. The output of the events generated is stored in a file. Once generated, the output file can be fed into the simulation program for further processing using a suitable interfacing program.

We have modified the event generator DECAY0 to generate events for the double beta decay process of ^{124}Sn in which we are interested. The required information

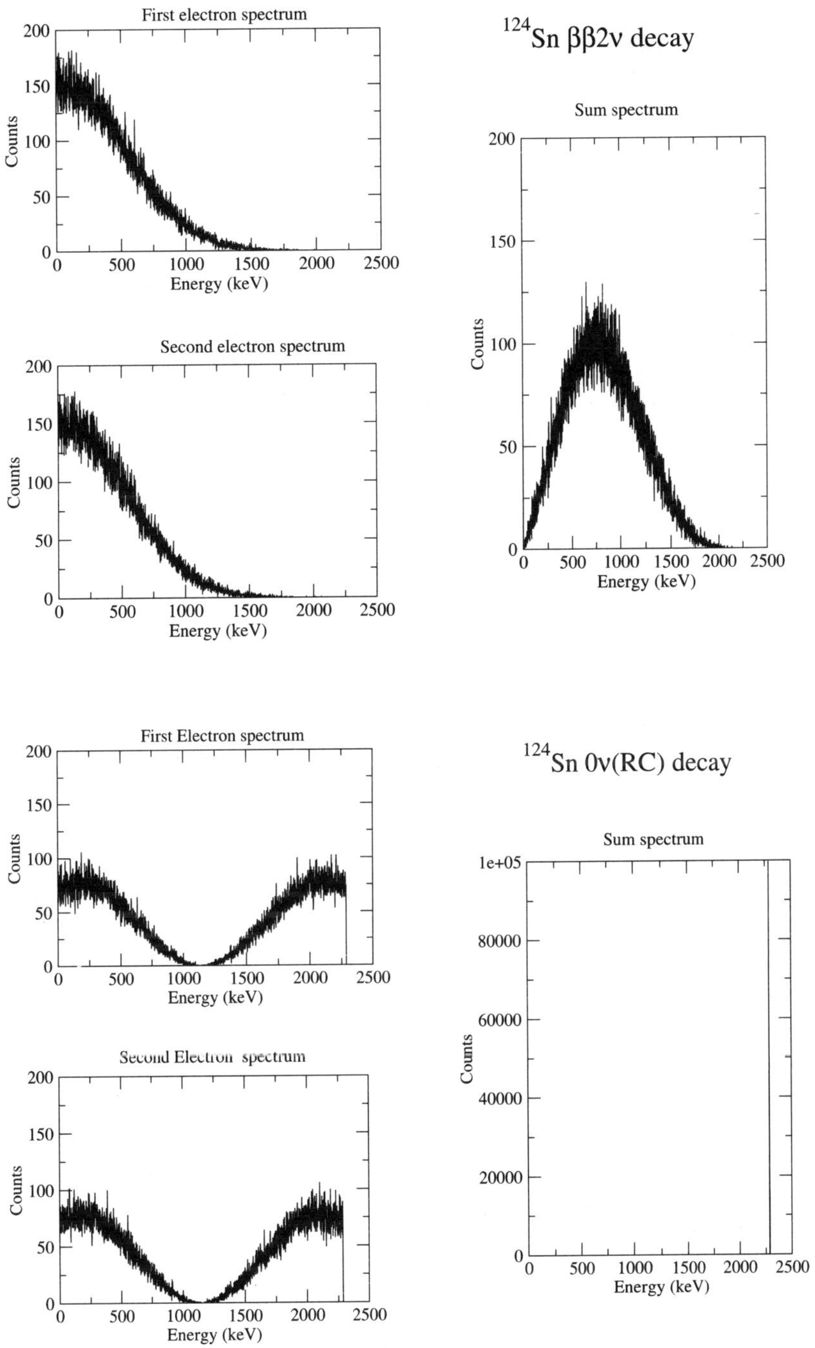

Figure 2. (Contd.)

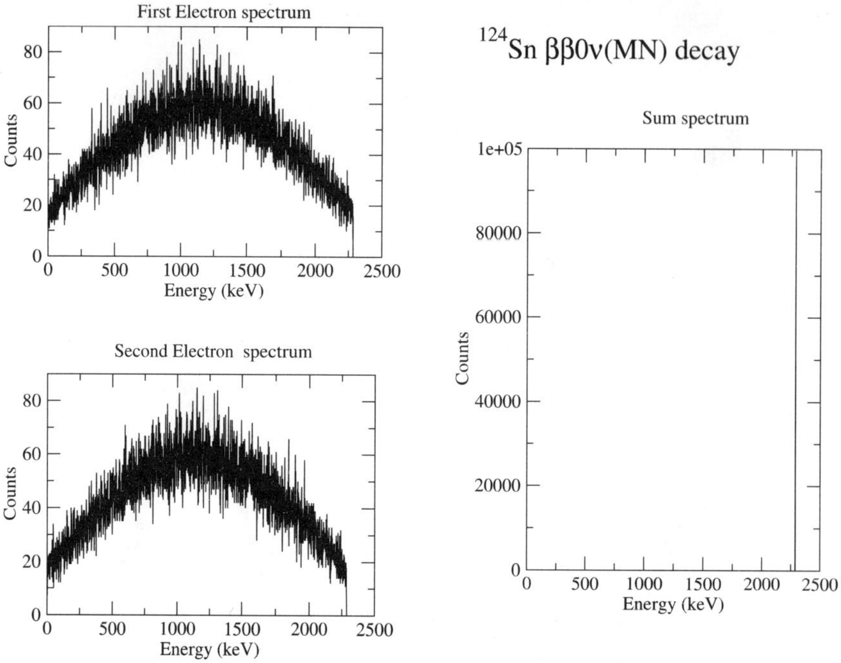

Figure 2. Single electron and the sum spectra for various modes and mechanisms in 2ν and 0ν double beta decay processes of ^{124}Sn. The decay events were generated for 0→0 transition using the modified event generator DECAY0. The "RC" and "MN" in the label represent the decay with right-handed current and neutrino mass mechanism, respectively

related to nuclear levels of Sn and and the daughter nucleus Te have been obtained from ENSDF database [8]. Typical spectra for varous modes of $2\nu\beta\beta$ and $0\nu\beta\beta$ decay processes of ^{124}Sn obtained from the modified event generator DECAY0 are shown in the Fig. 2.

4. Proposed Experiment

We have proposed to study double beta decay process of ^{124}Sn nucleus. One of the main considerations behind the selection of the candidate nucleus for the double beta decay process is a high Q value since the decay rate is proportional $\sim Q^5$ and the interference from natural radioactive background is less at higher energies. Moreover, as these events are expected to be rare, both source and detector have to be large in size. Therefore, large isotopic abundance of the isotopes are favourable. The Q value for the decay of ^{124}Sn is fairly large, i.e. Q = 2.28 MeV and the isotopic abundance of 5.8% is reasonable behind its selection. The experimental setup becomes simplified if the source itself can serve as a detector. One of the crucial parameter for any dector design is the energy resolution which is important for the precise measurement of sum energy of the two electrons emitted in $0\nu\beta\beta$ process. For this purpose, a bolometric detector is ideally suited. Tin becomes superconducting below 3.7 K and is nonmagnetic. Thus at temperatures below 100 mK the specific heat of Sn is very low, having only lattice contributions, and can be used as a bolometric detector. Details study about the possibilities that Tin can be used as a bolometric detector are discussed in the ref. [9].

We are in a process to develop different components of the simulation related to the geometry of the experimental setup and efficient read out of the signal in the Sn Bolometer.

References

[1] S. Fukuda *et al.* [Super-Kamiokande Collaboration], Phys. Rev. Lett. 86 (2002).

[2] Q.R. Ahmad *at al.* [SNO Collaboration], Phys. Rev. Lett. **87**, 073101 (2002); Q.R. Ahmad *at al.* [SNO Collaboration], Phys. Rev. Lett. **89**, 011301 (2002) and [arXiv:nucl-ex/0204008]; Q.R. Ahmad *at al.* [SNO Collaboration], Phys. Rev. Lett. **89**, 011302 (2002) and [arXiv:nucl-ex/0204009].

[3] K. Eguchi *et al.* [KamLAND Collaboration], Phys. Rev. Lett. **90**, 021802 (2003) and [arXiv:hep-ex/0212021].

[4] H.V. Klapdor-Kleingrothaus *et al.*, Mod. Phys. Lett. A **16**, 2409 (2001)

[5] H.V. Klapdor-Kleingrothaus, A. Dietz, I.V. Krivosheina, Part. Nucl. **110**, 57 (2002)

[6] The GEANT4 Collaborations, http://geant4.web.cern.ch/geant4/

[7] V. Tretyak, private communication.

[8] National Nuclear Data Centre, Brookhaven; http://www.nndc.bnl.gov

[9] V.Nanal *et al.*, see contribution in this proceeding.

NEUTRINOLESS DOUBLE BETA DECAY
Editors: V.K.B. Kota and U. Sarkar

NUCLEAR MODELS AND STATISTICAL SPECTROSCOPY FOR DOUBLE-BETA DECAY

V.K.B. Kota[1]

Physical Research Laboratory, Ahmedabad - 380 009, India

Abstract

Neutrinoless double-beta decay (NDBD) half-lives combined with the corresponding nuclear structure matrix elements (NSME), calculated using nuclear models/methods, give a value/bound for neutrino mass. In this paper first an overview of nuclear shell model and group theoretical models for the calculation of NSME for NDBD is given. Secondly, statistical spectroscopy method (SSM) for the NSME is introduced and discussed in detail. The SSM also allows for the study of double Gamow-Teller strength sums going beyond the standard $SU(4)$ description.

1. Introduction

Goeppert-Mayer [1] was the first in 1935, following the suggestion of Wigner, to predict that 2ν double beta-decay [2ν DBD or $\beta\beta(2\nu)$] half-lives could be $> 10^{20}$y. This

[1] **E-mail:vkbkota@prl.res.in**

process corresponds to (Z,A) $\longrightarrow$ (Z+2,A) + $e_1^- + e_2^- + \overline{\nu_{e_1}} + \overline{\nu_{e_2}}$. Soon after this, in 1937, following the fundamental suggestion of E. Majorana [2], Racah [3] pointed out the possibility of neutrinoless (or 0ν) DBD, hereafter called $\beta\beta(0\nu)$ or NDBD . This process corresponds to (Z,A) $\longrightarrow$ (Z+2,A) + $e_1^- + e_2^-$. Furry [4] in 1939, for the first time studied 0ν DBD half-lives. With these initial developments, large number of experimental and theoretical studies of both $\beta\beta(0\nu)$ and $\beta\beta(2\nu)$ are carried out. Some of the good candidates for 0ν (and also 2ν) DBD, with some experiments already done, are: ^{48}Ca, ^{76}Ge, ^{82}Se, ^{96}Zr, ^{100}Mo, ^{116}Cd, ^{124}Sn, ^{128}Te, ^{130}Te, ^{136}Xe, ^{150}Nd, ^{160}Gd and ^{238}U. Fundamental significance of NDBD, already discussed in the previous papers in this proceedings, is its experimental confirmation will tell us about lepton number violation in nature and that neutrino is a Majorana particle. In addition, the half-lives for $\beta\beta(0\nu)$ along with a good nuclear structure matrix elements calculation, i.e. $\left|M^{0\nu}(0^+)\right|^2$ in Eq. (1) ahead, will give a value or a bound on neutrino mass. So far only Klapdor et al [5] claimed to have evidence (at a confidence level of 4.2σ) for $\beta\beta(0\nu)$ in ^{76}Ge. However no evidence for $\beta\beta(0\nu)$ was found in the more recent ^{130}Te experiment [6]. Thus NDBD experiments at a much higher level of sensitivity are clearly called for. For reviews on DBD see [7, 8, 9] and for tabulation of experimental data and comparisons with theory (for NSME), see [10].

Our purpose in this paper is to give a general overview of nuclear models related to the shell model and also introduce a new method based on statistical nuclear spectroscopy, for calculating the NSME for NDBD and also for double Gamow-Teller strengths. Statistical spectroscopy [11, 12, 13] is based on the operation of Central Limit Theorems, unitary decompositions, two-body random matrix ensembles and quantum chaos. Here matrix elements are written in terms of traces of certain operators. It is important to stress that nuclear structure in general causes variation by a factor ~ 10 in the matrix elements and this implies a factor ~ 100 in half lives. Thus the need for reliable nuclear models/methods for NSME (for nuclei with $A < 60$ it is possible to carry out reliable nuclear shell model calculations, however, except for ^{48}Ca, all the nuclei involved in NDBD are with $A > 75$). We need nuclear models also for testing various approximations used in detailed models/methods and this is where group theoretical models are of great interest (as discussed in Section 3, some well developed group theoretical models are also used for calculating NSME). Although the major interest is in 0ν DBD, in general 2ν DBD is also studied as there is good data (for ~ 10 nuclei) for $\beta\beta(2\nu)$ half-lives and they can be used for calibrating nuclear models/methods that will be used for NDBD. Now we will give a preview.

In Section 2 given are expressions for 0ν and 2ν DBD half-lives and the corresponding transition operators. Section 3 gives a general discussion of nuclear models for NSME calculations for NDBD with particular reference to shell model and group theoretical models. In Section 4 introduced is statistical spectroscopy method (SSM) for calculating NSME for NDBD half-lives. Similarly Section 5 gives SSM for double Gamow-Teller (GT) strength sums (they are closely related to 2ν DBD). Finally Section 6 gives conclusions.

2. Double-Beta Decay Half-Lives and Transition Operators

2.1. 0ν *Double-Beta Decay*

Half-life for 0ν DBD, for the 0_i^+ ground state (gs) of a initial even-even nucleus decay to the 0_f^+ gs of the final even-even nucleus, is given by

$$\begin{gathered}\left[T_{1/2}^{0\nu}\left(0_i^+ \longrightarrow 0_f^+\right)\right]^{-1} = G^{0\nu}(\Delta E, Z)\ \left|M^{0\nu}(0^+)\right|^2\ \langle m_\nu\rangle^2\ , \\ M^{0\nu}(0^+) = M_{GT}^{0\nu}(0^+) - \frac{g_V^2}{g_A^2} M_F^{0\nu}(0^+)\end{gathered} \tag{1}$$

where ΔE is the Q-value (or the available energy $Q + 2m_e$ for the out going two electrons in NDBD) and $\langle m_\nu\rangle$ is effective neutrino mass. The $G^{0\nu}(\Delta E, Z)$ is accurately calculable phase space integral. Similarly g_A and g_V are the weak axial-vector and vector coupling constants ($g_A/g_V = 1$ to 1.254). The nuclear matrix elements M_{GT} and M_F are matrix elements of Gamow-Teller and Fermi like two-body operators respectively. The forms for them will follow from closure approximation which is well justified for NDBD. Then we have,

$$\begin{gathered}M^{0\nu}(0^+) = M_{GT}^{0\nu}(0^+) - \frac{g_V^2}{g_A^2} M_F^{0\nu}(0^+) = \left\langle 0_f^+ \ ||\ \mathscr{O}(2:0\nu)\ ||\ 0_i^+ \right\rangle\ , \\ \mathscr{O}(2:0\nu) = \sum_{\ell,k} \mathscr{H}(r_{\ell k}, \overline{E})\tau_\ell^+\tau_k^+(\sigma_\ell\sigma_k - \frac{g_V^2}{g_A^2})\ .\end{gathered} \tag{2}$$

Thus NDBD half-lives are generated by the two-body transition operator $\mathscr{O}(2:0\nu)$. The $\mathscr{H}(r_{\ell k}, \overline{E})$ in Eq. (2) is called 'neutrino potential'. Here $\overline{E} = \langle E_m\rangle - (M_i + M_f)/2$ where $\langle E_m\rangle$ is the average energy of the intermediate odd-odd nucleus and M_i and M_f are the masses of the initial and final even-even nuclei respectively. The form given by Eq. (2) for the nuclear matrix elements is justified *only if the exchange of the light Majorana neutrino is indeed the mechanism responsible for the* 0ν *DBD.* The effective neutrino potential, defining completely the two-body NDBD transition operator $\mathscr{O}(2:0\nu)$, is

$$\mathscr{H}(r_{\ell k}, \overline{E}) = \frac{R}{r_{\ell k}}\ e^{-\frac{3}{2}\frac{\overline{E}}{\hbar c} r_{\ell k}}\ \left[1 - e^{-\gamma_1 r_{\ell k}^2}\left(1 - \gamma_2\, r_{\ell k}^2\right)\right]\ . \tag{3}$$

Here $\gamma_1 = 1.1\,fm^{-2}$, $\gamma_2 = 0.68\,fm^{-2}$, R is nuclear radius in fm, $\overline{E}$ in MeV and $\hbar c = 197.327$ MeV fm. For more discussion on the validity of Eqs. (2, 3) see [7, 8, 14]. Thus, an experimental value of (bound on) $T_{1/2}^{0\nu}$ will give a value for (bound on) neutrino mass via Eq. (1) provided we know the value of the square of NSME $M^{0\nu}(0^+)$ of the NDBD two-body transition operator $\mathscr{O}(2:0\nu)$, connecting the gs of the initial and final even-even nuclei involved.

Let us say that for the nuclei under consideration, protons are in single particle (sp) orbits j^p and similarly neutrons in j^n. Using the usual assumption that the radial part of the sp states are those of the harmonic oscillator, the proton sp states are completely specified by $(\mathbf{n}^p, \ell^p, j^p)$ with $\mathbf{n}^p$ denoting oscillator radial quantum number so that for a oscillator shell $\mathscr{N}^p$, $2\mathbf{n}^p + \ell^p = \mathscr{N}^p$. Similarly the neutron sp states are $(\mathbf{n}^n, \ell^n, j^n)$. In terms of creation ($a^\dagger$) and annihilation ($a$) operators, normalized two-particle creation operator $A^J_\mu(j_1 j_2) = (1+\delta_{j_1 j_2})^{-1/2}(a^\dagger_{j_1} a^\dagger_{j_2})^J_\mu$ and then $A^J_\mu\,|0\rangle$ is a normalized two-particle state. With these the NDBD (in closure which is good) transition operator $\mathscr{O}(2:0\nu)$ in standard form, with V's denoting two-body matrix elements (TBME), is

$$\mathscr{O}(2:0\nu) = \sum_{j_1^p \geq j_2^p\,;\, j_3^n \geq j_4^n\,;\, J_0} V^{J_0}_{j_1^p j_2^p\,:\, j_3^n j_4^n}(0\nu) \sum_\mu A^{J_0}_\mu(j_1^p j_2^p) \left\{ A^{J_0}_\mu(j_3^n j_4^n) \right\}^\dagger . \tag{4}$$

Numerical values for the TBME $V(0\nu)$'s in Eq. (4) follow the definition of $\mathscr{O}(2:0\nu)$, transforming two-particle $j-j$ coupled states into $L-S$ coupled states and by applying Brody-Moshinsky brackets to transform the two-particle $L-S$ coupled states to states in relative and center-of-mass coordinates.

2.2. 2ν Double-Beta Decay

For 2ν DBD from even-even nuclei, the half-lives for 0^+ to 0^+ transitions are given by

$$\begin{gathered} \left[T^{2\nu}_{1/2}\left(0^+_i \longrightarrow 0^+_f\right) \right]^{-1} = G^{2\nu}(\Delta E, Z)\, \left|M^{2\nu}(0^+)\right|^2 , \\ M^{2\nu}(0^+) = M^{2\nu}_{GT}(0^+) \end{gathered} \tag{5}$$

where $G^{2\nu}(\Delta E, Z)$ is again an accurately calculable phase space integral. The nuclear $M^{2\nu}_{GT}$ matrix element is given by,

$$M^{2\nu}(0^+) = M^{2\nu}_{GT}(0^+) = \sum_m \frac{\left\langle 0^+_f \mid\mid \sigma\tau^+ \mid\mid 1^+_m \right\rangle \langle 1^+_m \mid\mid \sigma\tau^+ \mid\mid 0^+_i \rangle}{E_m - (M_i + M_f)/2} . \tag{6}$$

Here $|1^+_m\rangle$ are the virtual intermediate odd-odd nuclear states with energy E_m. In the closure approximation, though it is not very good for 2ν DBD, E_m is replaced by $\langle E_m \rangle$ and then carrying out the summation in Eq. (6) gives,

$$\begin{gathered} M^{2\nu}(0^+) = M^{2\nu}_{GT}(0^+) = \frac{\left\langle 0^+_f \mid \mathscr{O}(2:2\nu) \mid 0^+_i \right\rangle}{E_0 + \langle E_m \rangle - E_i} , \\ \mathscr{O}(2:2\nu) \longrightarrow \mathscr{O}^{DGT}(2) = \sqrt{3}\, \left[(\sigma\tau^+) \times (\sigma\tau^+) \right]^0 \end{gathered} \tag{7}$$

where the symbol $\times$ denotes the coupling of σ's and $E_0 = (M_i - M_f)/2 = \frac{1}{2}Q_{\beta\beta} + m_e c^2$. Thus, with closure, the $\beta\beta(2\nu)$ transition operator $\mathscr{O}(2:2\nu)$ is double Gamow-Teller (DGT) operator $\mathscr{O}^{DGT}(2)$ and it is a two-body operator. It is easy to write down the corresponding TBME, i.e. $V(DGT)$'s. To this end, first note that the $\sigma\tau^+$ operator can be written as [15]

$$\begin{gathered} (\sigma\tau^+)^1_\mu = \sum_{j_1^p,\, j_2^n} \varepsilon^1_{j_1^p,\, j_2^n} \left(a^\dagger_{j_1^p} \tilde{a}_{j_2^n} \right)^1_\mu , \\ \varepsilon^1_{(\mathbf{n}_1^p \ell_1^p j_1^p),(\mathbf{n}_2^n \ell_2^n j_2^n)} \overset{j_1^p = j_2^n}{\longrightarrow} \delta_{\ell_1^p \ell_2^n}\, \delta_{\mathbf{n}_1^p \mathbf{n}_2^n} \sqrt{\frac{2j_1^p+1}{3}}\, \frac{j_1^p(j_1^p+1) + 3/4 - \ell_1^p(\ell_1^p+1)}{\sqrt{j_1^p(j_1^p+1)}} , \\ \varepsilon^1_{(\mathbf{n}_1^p \ell_1^p j_1^p),(\mathbf{n}_2^n \ell_2^n j_2^n)} \overset{j_1^p \neq j_2^n}{\longrightarrow} \delta_{\ell_1^p \ell_2^n}\, \delta_{\mathbf{n}_1^p \mathbf{n}_2^n} (\pm) \sqrt{\frac{8\ell_1^p(\ell_1^p+1)}{3(2\ell_1^p+1)}} . \end{gathered} \tag{8}$$

In Eq. (8) we have $\pm$ for $j_1^p = \ell_1^p \mp \frac{1}{2}$ and $j_2^n = \ell_1^p \pm \frac{1}{2}$. Now applying the $9-j$ transformation to $(a^\dagger_{j_1^p}\tilde{a}_{j_2^n})^1 \cdot (a^\dagger_{j_3^p}\tilde{a}_{j_4^n})^1$, it is easy to see that $\mathcal{O}^{DGT}(2)$ can be written exactly in the form given by Eq. (4) with $V(0\nu)$'s replaced by $V(DGT)$'s. The TBME $V(DGT)$ are given by

$$\begin{array}{rcl} V^{J_0}_{j_1^p j_2^p : j_3^n j_4^n}(DGT) & = & v^{J_0}_{j_1^p j_2^p : j_3^n j_4^n} - (-1)^{j_1^p + j_2^p + J_0}(1-\delta_{j_1^p j_2^p})\, v^{J_0}_{j_2^p j_1^p : j_3^n j_4^n} \\ & & -(-1)^{j_3^n + j_4^n + J_0}(1-\delta_{j_3^n j_4^n})\, v^{J_0}_{j_1^p j_2^p : j_4^n j_3^n} \\ & & +(-1)^{j_1^p + j_2^p + j_3^n + j_4^n}(1-\delta_{j_1^p j_2^p})(1-\delta_{j_3^n j_4^n})\, v^{J_0}_{j_2^p j_1^p : j_4^n j_3^n} \end{array} \tag{9}$$

where $v^{J_0}_{j_1^p j_2^p : j_3^n j_4^n}$ are,

$$\begin{array}{rcl} v^{J_0}_{j_1^p j_2^p : j_3^n j_4^n} & = & -3\, \varepsilon^1_{j_1^p, j_3^n}\, \varepsilon^1_{j_2^p, j_4^n} \left\{ \begin{array}{ccc} j_1^p & j_3^n & 1 \\ j_4^n & j_2^p & J_0 \end{array} \right\} \\ & & \times\, (-1)^{j_2^p + j_3^n + J_0} \sqrt{\left(1+\delta_{j_1^p j_2^p}\right)\left(1+\delta_{j_3^n j_4^n}\right)}\,. \end{array} \tag{10}$$

Before turning to nuclear models for DBD, let us mention that a much better alternative to closure for $\beta\beta(2\nu)$, is the so-called summation method. Here used is the identity $(E_0 + E_m - E_i)^{-1} = \int_0^\infty d\omega e^{-\omega(E_0+E_m-E_i)}$ and this gives (with H denoting Hamiltonian),

$$M^{2\nu}(0^+) = \int_0^\infty d\omega\, e^{-\omega E_0} \left\langle 0_f^+ \mid (\sigma\tau^+) e^{-\omega H} (\sigma\tau^+) e^{\omega H} \mid 0_i^+ \right\rangle . \tag{11}$$

Expanding $e^{-\omega H}(\sigma\tau^+)e^{\omega H}$ into multiple commutators involving H and $(\sigma\tau^+)$, one can write $M^{2\nu}(0^+)$ as a infinite series where each term in the series is a matrix element of a multiple commutator. This series takes a particularly simple form if the two-body part of the nuclear effective interaction H commutes with the GT operator $(\sigma\tau^+)$ so that in evaluating the effects of the denominator in Eq. (6), only the one-body part of H contributes [16]. Then symbolically, with $H = \sum_k \varepsilon_k a_k^\dagger a_k + \text{constant}$, and $(\sigma\tau^+) = \sum_{p,n} f(p,n) a_p^\dagger a_n$, we obtain

$$M^{2\nu}(0^+) = \sum_{p,n} \sum_{p',n'} f(p,n) f(p',n') \frac{\left\langle 0_f^+ \mid (a_p^\dagger a_n)(a_{p'}^\dagger a_{n'}) \mid 0_i^+ \right\rangle}{E_0 + \varepsilon_{p'} - \varepsilon_{n'}} . \tag{12}$$

Eq. (12) is used in many applications; see for example [17].

3. Nuclear Models for Neutrinoless Double-Beta Decay Nuclear Structure Matrix Elements

Klapdor, Faessler, Suhonen, Civitarese and others used quite extensively the quasi-particle random phase approximation (QRPA) and its various modified versions for DBD calculations; see [7, 10, 18] and references therein. For example, as listed in [9], there are 22 NSME calculations for NDBD for ^{76}Ge using QRPA. As large literature is available, we will not discuss this method any further. Another class of models are those based on mean-field methods. For example for well deformed nuclei it is possible to use a single Hartree-Fock intrinsic state with angular momentum projection, i.e.

projected Hartree-Fock (PHF) or by including pairing correlations via the projected Hartree-Fock-Bogoliubov (PHFB) method where a single HFB intrinsic state is used. The paper by Rath in this proceedings gives the results of PHFB for NDBD (and 2ν DBD). By including band mixing with PHF gives the so called deformed shell model (DSM) and similarly band mixing with PHFB gives Sun's [19] projected shell model (PSM). Sahu's paper in this proceedings gives some results for DBD obtained using DSM. It is important to stress that a nuclear model/method becomes more trustworthy if it can reproduce more observables (they can fix free parameters if any in the model/method). Further they will become more relevant if the observables resemble more closely $\beta\beta(0\nu)$ decay. Now we will discuss in some detail applications of shell model and group theoretical models for NDBD and in Section 4 ahead a new method based on statistical nuclear spectroscopy is introduced.

3.1. Shell Model for NDBD

Exact shell model is best for NSME calculations. However with shell model the questions to be addressed are: (i) choice of valence space (truncation); (ii) effective interaction, (iii) construction of H matrix and (iv) diagonalization. With large shell model codes, in particular due to Caurier et al [20], (iii) and (iv) are pushed to very large dimensions but still for $A > 75$ nuclei, it is necessary to truncate the shell model spaces. For example for ^{124}Sn with r_4h (defined ahead) space for neutrons, dimension $d(J=0) \sim 10^6$. Inclusion of $2p-2h$ excitations for protons from $^1g_{9/2}$, the dimension grows to $d \sim 10^{10}$. Although one can use Z=50 core for ^{124}Sn and N=82 core for ^{136}Xe, particle-hole excitations from core should be included because of the structure of DBD transition operators (they will connect configurations with protons and neutrons in sp orbits that are spin-orbit partners). With regard to effective interactions, there is now good progress with the monopole correction to realistic interactions due to Zuker et al [21] and with the advances in effective interaction theory due to Hjorth-Jensen et al [22]. Let us denote $(^1f_{7/2}, {}^2p_{3/2}, {}^2p_{1/2}, {}^1f_{5/2})$ by fp, $(^2p_{3/2}, {}^2p_{1/2}, {}^1f_{5/2})$ by r_3, $^1g_{9/2}$ by g, $(^1g_{7/2}, {}^2d_{5/2}, {}^2d_{3/2}, {}^3s_{1/2})$ by r_4, $^1h_{11/2}$ by h, $(^2f_{7/2}, {}^1h_{9/2}, {}^3p_{3/2}, {}^2f_{5/2}, {}^3p_{1/2})$ by r_5 and $^1i_{13/2}$ by i. Till now best shell model calculations for NDBD using shell model are due to Caurier et al with results for 7 nuclei reported; see the first reference in [20]. These nuclei and the shell model spaces employed for them are: ^{48}Ca (fp shell); ^{76}Ge, ^{82}Se (r_3g); ^{124}Sn, ^{128}Te, ^{130}Te, ^{136}Xe (r_4h space). Assuming $T^{0\nu}_{1/2} = 10^{25}$yr, the predicted values for $\langle m_\nu \rangle$ are 0.85, 0.9, 0.42, 0.45, 1.92, 0.35, and 0.41 eV respectively. Besides these, in the second reference in [20], calculations for ^{110}Pd and ^{116}Cd are also reported with $\langle m_\nu \rangle = 0.67$ and 0.24 eV respectively. Finally we mention that, for very large shell model spaces, shell model Monte-Carlo (SMMC) method is an alternative but so far there are only limited studies of DBD using SMMC [23].

It is useful to mention that large shell model codes will be very useful in testing the statistical spectroscopy method, for NDBD, introduced ahead in Section 4, i.e. the structures given by Eq. (16) for NSME, Eq. (22) and its further extensions for double GT strength sums, and the expression for ζ in Eq. (20).

3.2. Group Theoretical Models for NDBD

Simplest group theoretical model is the $SU(2)$ pairing model for identical nucleons in single or several-j shells [24]. The $SU(2)$ algebra is generated by the pair creation operator $S_+ = \sum_j \beta_j \sqrt{2j+1}(a_j^\dagger a_j^\dagger)^0$, $S_- = (S_+)^\dagger$ and $S_0 = (\hat{n} - \Omega_0)/2$; here $\beta_j = \pm 1$, $\Omega_0 = [\sum_j (2j+1)]/2$ and $\hat{n}$ is number operator. For a given number m of identical nucleons, the $SU(2)$ algebra generates the states $|m, v, \alpha\rangle$ which are eigenstates of the pairing Hamiltonian $H_p = -GS_+S_-$ and v is the well known (generalized) seniority quantum number. Truncation based on v has been used for example in the $p-SU(3)$ model for NDBD discussed ahead. For nucleons in single or several-j shells (i.e. in $j-j$ coupling), the pairing algebra is $SO(5)$ generated by the isovector pair creation operators $P_\mu^\dagger = \sum_j \beta_j \sqrt{2j+1}(a^\dagger_{j\frac{1}{2}} a^\dagger_{j\frac{1}{2}})^{J_0=0,T_0=1}_{0,\mu}$, pair annihilation operators P_μ, isospin generators T_μ and $(\hat{n} - \Omega_1)/2$; $\Omega_1 = \sum_j (2j+1)$ and $\beta_j = \pm 1$. The isovector pairing Hamiltonian $H_p = -GP^\dagger \cdot P$ generates the $SO(5)$ states $|m, v, t, T, \alpha\rangle$ where t is reduced isospin. $SO(5)$ algebra has been well developed by Hecht [25] and it is used by Engel et al [26] to understand the quality of QPRA and effects of shell model truncation for NDBD. Main problem with the $SO(5)$ model is that the isoscalar pairing operators are not in this algebra. The $SO(8)$ pairing model (see [27] and references therein) includes, in $L-S$ coupling both isoscalar (D) and isovector (P) pairing. The $SO(8)$ algebra is generated by the 28 operators

$$\begin{aligned} P_\mu^\dagger &= \textstyle\sum_\ell \beta_\ell \sqrt{(2\ell+1)/2}\, (a^\dagger_{\ell,\frac{1}{2},\frac{1}{2}} a^\dagger_{\ell,\frac{1}{2},\frac{1}{2}})^{L_0=0,S_0=0,T_0=1}_{0,0,\mu} , \\ D_\mu^\dagger &= \textstyle\sum_\ell \beta_\ell \sqrt{(2\ell+1)/2}\, (a^\dagger_{\ell,\frac{1}{2},\frac{1}{2}} a^\dagger_{\ell,\frac{1}{2},\frac{1}{2}})^{L_0=0,S_0=1,T_0=0}_{0,\mu,0} , \\ & P\mu,\ D\mu,\ S_\mu,\ T_\mu,\ (\sigma\tau)^{1,1}_{\mu,\mu'},\ (\tfrac{\hat{n}}{2} - \Omega) \end{aligned} \tag{13}$$

where $\beta_\ell = \pm 1$ and $\Omega = \sum_\ell (2\ell+1)$. The isoscalar plus isovector pairing Hamiltonian

$$H_p = -(1-x)\sum_\mu P_\mu^\dagger P_\mu - (1+x)\sum_\mu D_\mu^\dagger D_\mu \tag{14}$$

for $x = 0$ gives a $SO(8)$ subgroup chain with Wigner's spin-isospin $SU(4)$ subalgebra. Similarly with $x = 1, -1$ there are subgroup chains with $SO(5)$. Thus H_p with $x = 0, 1, -1$ is solvable. Major problem with the $SO(8)$ model is that there are very few results for the Wigner-Racah algebra of the $SO(8)$ subgroup chains. Engel et al [14, 28] used the $SO(8)$ model for understanding certain aspects of QRPA for DBD. In addition to the pairing models, interacting boson model was also used for DBD [29].

Most extensively used group theoretical model for NDBD (also for 2ν DBD) NSME calculations is the pseudo SU(3) [hereafter called $p-SU(3)$] model; see Refs. [17, 30] and references therein. The $p-SU(3)$ model is usually used for heavy deformed nuclei with valance protons and neutrons in different oscillator shells. Then one has the situation with protons (similarly neutrons) occupying both natural and intruder parity orbitals. For example in ^{150}Nd, with Z=50, N=82 as closed shells, there are $m_p = 10$ protons in r_4h space (r_4 orbits being of +ve parity and the intruder $h={}^1h_{11/2}$ orbit being of −ve parity) and $m_n = 8$ neutrons in r_5i space (r_5 orbits being of −ve parity and the intruder $i={}^1i_{13/2}$ orbit being of +ve parity). As the r_4 orbitals map to pseudo oscillator $\widetilde{\mathcal{N}} = 3$ shell and similarly r_5 to $\widetilde{\mathcal{N}} = 4$, it is possible to have Elliott's SU(3), i.e. a pseudo $SU(3)$ algebra for proton and neutrons in the natural parity orbits. Note that $j = \tilde{\ell} + \tilde{s}$ with $\tilde{s} = 1/2$ being the pseudo-spin and for a $\widetilde{\mathscr{N}}$ shell the $\tilde{\ell} \pm 1/2$ are nearly degenerate (see Fig. 1 for $\widetilde{\mathscr{N}} = 3$ example). Now starting with the mapping of Nilsson orbits to pseudo-Nilsson orbits with intruder orbit Nilsson levels in between (see Fig. 1 for an example), we can distribute, at a given deformation of the nucleus ^{150}Nd in the present discussion, protons and neutrons in r_4h and r_5i space orbits respectively. Let us denote m_p^{na} and m_p^{int} as the number of protons in natural parity pseudo-Nilsson orbits $[\widetilde{\mathscr{N}}, n, \Lambda]$ and the intruder parity orbital ($h_{11/2}$ in Fig. 1) respectively. Subject to Pauli principle these numbers can be determined for the lowest configuration. Assuming that the pseudo spin $\tilde{S}_p = 0$ for m_p^{na} protons, from the occupied $[\widetilde{\mathscr{N}}, n, \Lambda]$ orbits one can identify the appropriate pseudo-$SU(3)$ quantum numbers (λ^p, μ^p). Then, applying the pairing $SU(2)$ to m_p^{int} and assuming the seniority $v^{p:int} = 0$, we have a truncated shell model space for protons as given by $\left|m_p^{na}, \tilde{S}_p = 0, (\lambda^p, \mu^p), \tilde{L}_p^{na}; m_p^{int}, v^{p:int} = 0, L_{int} = 0; L_p = \tilde{L}_p^{na}\right\rangle$. Similarly, $\left|m_n^{na}, \tilde{S}_n = 0, (\lambda^n, \mu^n), \tilde{L}_n^{na}; m_n^{int}, v^{n:int} = 0, L_{int} = 0; L_n = \tilde{L}_n^{na}\right\rangle$ are the states for neutrons. In the resulting proton-neutron space, using the $SU(3)$ algebra, one can calculate NSME for NDBD [it is also possible to expand the $p-SU(3)$ model space]. Hirsch et al. applied the $p-SU(3)$ model to a number of heavy nuclei and for example [30] the predicted NDBD half-lives, assuming $\langle m_\nu \rangle = 1$eV, for ^{146}Nd, ^{148}Nd, ^{150}Nd, ^{186}W, ^{192}Os and ^{238}U are 1.18×10^{28}yr, 6.75×10^{24}yr, 1.05×10^{24}yr, 5.13×10^{25}yr, 3.28×10^{26}yr and 1.03×10^{24}yr respectively. A test of the model is from the observed 2ν DBD half-lives 9×10^{18}yr for ^{150}Nd and 2×10^{21}yr for ^{238}U and they are in good agreement with the $p-SU(3)$ results [30].

4. Statistical Spectroscopy Method for Neutrinoless Double-Beta Decay Nuclear Structure Matrix Elements

Following Eqs. (1,2) for NDBD half-lives, the corresponding NSME $\left|M^{0\nu}\right|^2$ can be viewed as transition strengths generated by the two-body transition operator $\mathscr{O}(2:0\nu)$. Therefore statistical spectroscopy theory [11, 13] for transition strengths can be applied here. Similarly in closure approximation $\left|M^{2\nu}\right|^2$ is a transition strength generated by $\mathscr{O}(2:2\nu)$ operator given in Eq. (7) with TBME defined by Eqs. (9,10). As closure is not good for 2νDBD, we will not consider this process in the following. We will now discuss statistical spectroscopy method (SSM) for NDBD and the problems that remain to be solved for obtaining numerical values for the half-lives.

Let us consider shell model spherical orbits with angular momenta $j_1^p, j_2^p, \ldots, j_r^p$ with m_p protons distributed over them and similarly m_n neutrons over $j_1^n, j_2^n, \ldots, j_s^n$ orbits. Then the proton configurations $\widetilde{m_p} = [m_p^1, m_p^2, \ldots, m_p^r]$ where m_p^i is number of protons in the orbit j_i^p and $\sum_{i=1}^r m_p^i = m_p$. Similarly the neutron configurations $\widetilde{m_n} = [m_n^1, m_n^2, \ldots, m_n^s]$ where m_n^i is number of neutrons in the orbit j_i^n and $\sum_{i=1}^s m_n^i = m_n$. With these, $(\widetilde{m_p}, \widetilde{m_n})$'s denote proton-neutron configurations. The nuclear effective Hamiltonian is one plus two-body, $H = \mathbf{h} + \mathbf{V}$ and we assume that the one-body part $\mathbf{h}$ includes the mean-field producing part of the two-body part. Thus [12], $\mathbf{V}$ is the irreducible two-body part of H. The state density $I^H(E)$, with $\langle\langle - - \rangle\rangle$ denoting trace, can be written as a sum of partial densities defined over $(\widetilde{m_p}, \widetilde{m_n})$, i.e. $I^{(m_p, m_n)}(E) = \langle\langle \delta(H-E) \rangle\rangle^{(m_p, m_n)} = \sum_{(\widetilde{m_p}, \widetilde{m_n})} \langle\langle \delta(H-E) \rangle\rangle^{(\widetilde{m_p}, \widetilde{m_n})} =$

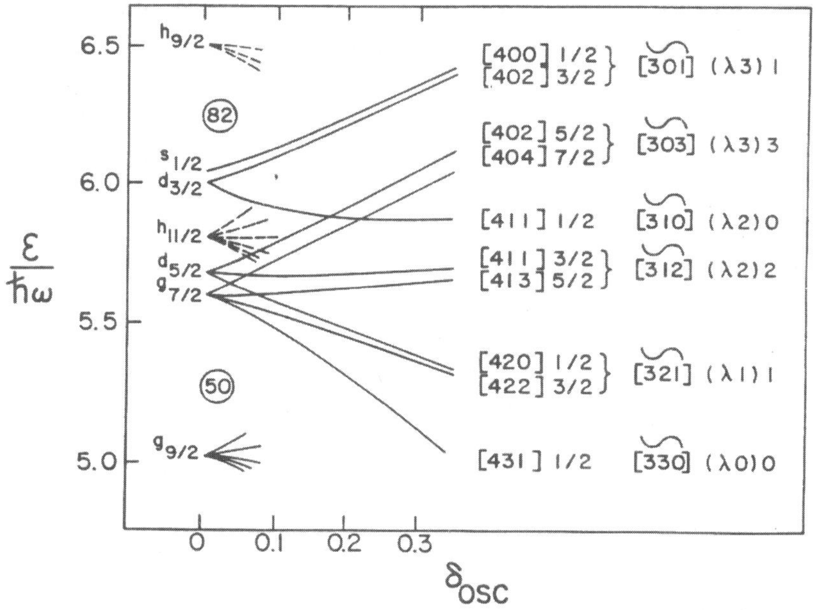

Figure 1. Nilsson diagram for $\widetilde{\mathscr{N}}=3$ shell. Pseudo Nilsson quantum numbers $[\widetilde{\mathscr{N},n,\Lambda}]$ for the natural parity orbits are shown in the figure. For avoiding over crowding, Nilsson levels from the intruder $h_{11/2}$ orbit (also from the lower $g_{9/2}$ and the higher $h_{9/2}$ orbit) are not shown in full in the figure. The figure also gives the IBFM $SU(3)\otimes SU(2)$ limit quantum numbers $(\lambda,\mu)K=\widetilde{\Lambda}$. See Ref. [31] for details and the figure is taken from the first reference in [31].

$\sum_{(\widetilde{m_p},\widetilde{m_n})} I^{(\widetilde{m_p},\widetilde{m_n})}(E)=\sum_{(\widetilde{m_p},\widetilde{m_n})} d(\widetilde{m_p},\widetilde{m_n})\,\rho^{(\widetilde{m_p},\widetilde{m_n})}(E)$. Here $d(\widetilde{m_p},\widetilde{m_n})$ is dimension and $\rho^{(\widetilde{m_p},\widetilde{m_n})}(E)$ is normalized to unity. With SS generated by the embedded GOE of one plus two-body interactions, i.e. EGOE(1+2) operating in the Gaussian domain (see [11]), locally will give, [as partial densities are nothing but sums over the strength functions],

$$I^{(m_p,m_n)}(E)=\sum_{(\widetilde{m_p},\widetilde{m_n})} I_{\mathscr{G}}^{(\widetilde{m_p},\widetilde{m_n})}(E)\,. \tag{15}$$

In Eq. (15) $\mathscr{G}$ denotes Gaussian. The Gaussian partial densities are defined by the centroids $E_c(\widetilde{m_p},\widetilde{m_n})=\langle H\rangle^{(\widetilde{m_p},\widetilde{m_n})}\sim\langle\mathbf{h}\rangle^{(\widetilde{m_p},\widetilde{m_n})}$ and variances $\sigma^2(\widetilde{m_p},\widetilde{m_n})=\left\langle H^2\right\rangle^{(\widetilde{m_p},\widetilde{m_n})}-[E_c(\widetilde{m_p},\widetilde{m_n})]^2\sim\left\langle \mathbf{V}^2\right\rangle^{(\widetilde{m_p},\widetilde{m_n})}$ and, as they are traces over $(\widetilde{m_p},\widetilde{m_n})$ spaces, they can be calculated without recourse to H matrix construction. Propagation equations for them, in terms of the sp energies defining **h** and TBME defining **V**, are easy to write down; see for example the first reference in [11]. Unitary group tensorial structure of **h** and **V** gives further simplifications of Eq. (15); see [12, 32] for details and also for applications in level density data analysis.

EGOE theory for transition strengths follows from the recognition that transition strength density $I_{\mathscr{O}}(E_i,E_f)=I(E_f)|\left\langle E_f\mid\mathscr{O}\mid E_i\right\rangle|^2 I(E_i)$ is a convolution of the corresponding transition strength density due to the mean-field part **h** with a spreading bivariate Gaussian $\rho_{biv-\mathscr{G}:\mathscr{O}(2:0\nu):\mathbf{V}}$ due to the interaction **V**. This gives [11, 13], for the square of the $\mathscr{O}(2:0\nu)$ matrix elements, with $\left|0_i^+\right\rangle = |E_i\rangle$ and $\left|0_f^+\right\rangle = |E_f\rangle$ where E's are energies,

$$
\begin{array}{l}
\left|\langle E_f \mid \mathscr{O}(2:0\nu) \mid E_i\rangle\right|^2 = \\
\displaystyle\sum_{(\widetilde{m_p},\widetilde{m_n})_i,(\widetilde{m_p},\widetilde{m_n})_f} \frac{I_{\mathscr{G}}^{(\widetilde{m_p},\widetilde{m_n})_i}(E_i)\, I_{\mathscr{G}}^{(\widetilde{m_p},\widetilde{m_n})_f}(E_f)}{I^{(m_p^i,m_n^i)}(E_i)\, I^{(m_p^f,m_n^f)_f}(E_f)} \\
\times \left|\langle(\widetilde{m_p},\widetilde{m_n})_f \mid \mathscr{O}(2:0\nu) \mid (\widetilde{m_p},\widetilde{m_n})_i\rangle\right|^2 \\
\times \dfrac{\rho_{biv-\mathscr{G}:\mathscr{O}(2:0\nu):\mathbf{V}}^{(\widetilde{m_p},\widetilde{m_n})_i,(\widetilde{m_p},\widetilde{m_n})_f}(E_i,E_f\,;\,E_c^i,E_c^f,\sigma_i,\sigma_f,\zeta)}{\rho_{\mathscr{G}}^{(\widetilde{m_p},\widetilde{m_n})_i}\;\rho_{\mathscr{G}}^{(\widetilde{m_p},\widetilde{m_n})_f}} \,.
\end{array}
\tag{16}
$$

In Eq. (16), $\left|\langle(\widetilde{m_p},\widetilde{m_n})_f \mid \mathscr{O}(2:0\nu) \mid (\widetilde{m_p},\widetilde{m_n})_f\rangle\right|^2$ is the mean square matrix element of $\mathscr{O}(2:0\nu)$ connecting $(\widetilde{m_p},\widetilde{m_n})_i$ and $(\widetilde{m_p},\widetilde{m_n})_f$ configurations,

$$
\begin{array}{l}
\left|\langle(\widetilde{m_p},\widetilde{m_n})_f \mid \mathscr{O}(2:0\nu) \mid (\widetilde{m_p},\widetilde{m_n})_i\rangle\right|^2 = \dfrac{\left|\langle\langle(\widetilde{m_p},\widetilde{m_n})_f \mid \mathscr{O}(2:0\nu) \mid (\widetilde{m_p},\widetilde{m_n})_i\rangle\rangle\right|^2}{d[(\widetilde{m_p},\widetilde{m_n})_i]\, d[(\widetilde{m_p},\widetilde{m_n})_f]} \,, \\
\left|\langle\langle(\widetilde{m_p},\widetilde{m_n})_f \mid \mathscr{O}(2:0\nu) \mid (\widetilde{m_p},\widetilde{m_n})_i\rangle\rangle\right|^2 = \\
\displaystyle\sum_{\alpha\in(\widetilde{m_p},\widetilde{m_n})_i,\, \beta\in(\widetilde{m_p},\widetilde{m_n})_f} \left|\langle(\widetilde{m_p},\widetilde{m_n})_f\, \beta \mid \mathscr{O}(2:0\nu) \mid (\widetilde{m_p},\widetilde{m_n})_i\, \alpha\rangle\right|^2 .
\end{array}
\tag{17}
$$

For later reference it is also useful define,

$$
\begin{array}{l}
\left\langle[\mathscr{O}(2:0\nu)]^\dagger \mathscr{O}(2:0\nu)\right\rangle^{(\widetilde{m_p},\widetilde{m_n})_i} = \{d[(\widetilde{m_p},\widetilde{m_n})_i]\}^{-1} \\
\times \displaystyle\sum_{(\widetilde{m_p},\widetilde{m_n})_f} \left|\langle\langle(\widetilde{m_p},\widetilde{m_n})_f \mid \mathscr{O}(2:0\nu) \mid (\widetilde{m_p},\widetilde{m_n})_i\rangle\rangle\right|^2 , \\
\left\langle\mathscr{O}(2:0\nu)[\mathscr{O}(2:0\nu)]^\dagger\right\rangle^{(\widetilde{m_p},\widetilde{m_n})_f} = \{d[(\widetilde{m_p},\widetilde{m_n})_f]\}^{-1} \\
\times \displaystyle\sum_{(\widetilde{m_p},\widetilde{m_n})_i} \left|\langle\langle(\widetilde{m_p},\widetilde{m_n})_f \mid \mathscr{O}(2:0\nu) \mid (\widetilde{m_p},\widetilde{m_n})_i\rangle\rangle\right|^2 .
\end{array}
\tag{18}
$$

Following [13], the trace forms for the two centroids E_c^i and E_c^f and the two variances σ_i^2 and σ_f^2 of the marginal densities of the bivariate Gaussian $\rho_{biv-\mathscr{G}:\mathscr{O}(2:0\nu):\mathbf{V}}$ are,

$$
\begin{array}{l}
E_c^i = \left\langle[\mathscr{O}(2:0\nu)]^\dagger \mathscr{O}(2:0\nu) H\right\rangle^{(\widetilde{m_p},\widetilde{m_n})_i} / \left\langle[\mathscr{O}(2:0\nu)]^\dagger \mathscr{O}(2:0\nu)\right\rangle^{(\widetilde{m_p},\widetilde{m_n})_i} , \\
E_c^f = \left\langle\mathscr{O}(2:0\nu)[\mathscr{O}(2:0\nu)]^\dagger H\right\rangle^{(\widetilde{m_p},\widetilde{m_n})_f} / \left\langle\mathscr{O}(2:0\nu)[\mathscr{O}(2:0\nu)]^\dagger\right\rangle^{(\widetilde{m_p},\widetilde{m_n})_f} , \\
\sigma_i^2 = \left\langle[\mathscr{O}(2:0\nu)]^\dagger \mathscr{O}(2:0\nu) \mathbf{V}^2\right\rangle^{(\widetilde{m_p},\widetilde{m_n})_i} / \left\langle[\mathscr{O}(2:0\nu)]^\dagger \mathscr{O}(2:0\nu)\right\rangle^{(\widetilde{m_p},\widetilde{m_n})_i} , \\
\sigma_f^2 = \left\langle\mathscr{O}(2:0\nu)[\mathscr{O}(2:0\nu)]^\dagger \mathbf{V}^2\right\rangle^{(\widetilde{m_p},\widetilde{m_n})_f} / \left\langle\mathscr{O}(2:0\nu)[\mathscr{O}(2:0\nu)]^\dagger\right\rangle^{(\widetilde{m_p},\widetilde{m_n})_f} .
\end{array}
\tag{19}
$$

Although the general structure and importance of the bivariate correlation coefficient ζ in Eq. (16) is well understood, an expression for ζ in terms of traces over $(\widetilde{m_p},\widetilde{m_n})_i$ and $(\widetilde{m_p},\widetilde{m_n})_f$ spaces is not yet available. Then the standard approximation, for completing the theory, is

$$
\zeta \sim \frac{\left\langle[\mathscr{O}(2:0\nu)]^\dagger \mathbf{V} \mathscr{O}(2:0\nu) \mathbf{V}\right\rangle^{m_p^i,m_n^i}}{\left\langle[\mathscr{O}(2:0\nu)]^\dagger \mathscr{O}(2:0\nu)\right\rangle^{m_p^i,m_n^i} \left\langle\mathbf{V}\mathbf{V}\right\rangle^{m_p^i,m_n^i}} \,.
\tag{20}
$$

To proceed further with SSM, propagation formulas for the traces in Eq. (17, 19, 20) are needed. Only for the trace in Eq. (17) it is easy to write the propagation formula in terms of the TBME $V^{J_0}_{--}(0\nu)$ defined by Eq. (4) and the result is (derived using Eq. (E.12) in the first reference in [11]),

$$\begin{aligned}
&\left|\langle\langle(\widetilde{m_p},\widetilde{m_n})_f \mid \mathscr{O}(2:0\nu) \mid (\widetilde{m_p},\widetilde{m_n})_i\rangle\rangle\right|^2 = \{d[(\widetilde{m_p},\widetilde{m_n})_i]\} \\
&\times \sum_{\alpha,\beta,\gamma,\delta} \frac{m^i_n(\alpha)[m^i_n(\beta)-\delta_{\alpha\beta}][N_p(\gamma)-m^i_p(\gamma)][N_p(\delta)-m^i_p(\delta)-\delta_{\gamma\delta}]}{N_n(\alpha)[N_n(\beta)-\delta_{\alpha\beta}]N_p(\gamma)[N_p(\delta)-\delta_{\gamma\delta}]} \\
&\times \sum_{J_0}\left[V^{J_0}_{\gamma^p\delta^p\alpha^n\beta^n}(0\nu)\right]^2(2J_0+1)\,, \\
&(\widetilde{m_p},\widetilde{m_n})_f = (\widetilde{m_p},\widetilde{m_n})_i \times \left(1^+_{\gamma_p}1^+_{\delta_p}1_{\alpha_n}1_{\beta_n}\right)\,.
\end{aligned} \tag{21}$$

Note that in Eq. (21) the final configuration is defined by removing one neutron from orbit α and another from β and then adding one proton in orbit γ and another in orbit δ. Also, $N_p(\alpha)$ is the degeneracy of the proton orbit α and similarly $N_n(\alpha)$ for the neutron orbit α. In the limit $\rho_{biv}/\rho\rho = 1$ in Eq. (16), substituting the result of Eq. (21) in Eq. (16) gives the NSME for NDBD. However, this zero-th order approximation will not be good as in general it is expected that $\zeta > 0.5$. Therefore for proper SSM results it is essential to use Eq. (16) in full. For this, we need the traces in Eqs. (19, 20); propagation formulas for these complicated traces are not yet available (they require extension of the methods given in [33]). These propagation formulas are being derived and in the near future SSM results for NDBD in various nuclei will be reported; SSM is best suited for heavy nuclei with many valance nucleons.

5. Statistical Spectroscopy Method for Double GT Strength Sums

Total strength generated by double GT operator acting on the 0^+_i ground state of a even-even nucleus is a quantity of interest in DBD studies and this is closely related to 2ν DBD. This is simply given by the expectation value of the operator $[\mathscr{O}^{DGT}(2)]^\dagger\mathscr{O}^{DGT}(2)$ in the ground state of the initial nucleus. So far using only the Wigner's spin-isospin $SU_{ST}(4)$ symmetry, Vogel et al and Muto [34] studied double GT strength sums and it is observed that the ground to ground double GT transition exhausts only a small fraction of the total strength sum, therefore $M^{2\nu}_{GT}$ is very sensitive to the nuclear structure details. Using SSM, it is possible to go beyond the $SU(4)$ theory for double GT strength sums. In the lowest approximation, SSM gives

$$\begin{aligned}
&\left\langle[\mathscr{O}^{DGT}(2)]^\dagger\mathscr{O}^{DGT}(2)\right\rangle^{E_i} = \\
&\sum_{(\widetilde{m_p},\widetilde{m_n})_i} \frac{I^{(\widetilde{m_p},\widetilde{m_n})_i}_{\mathscr{G}}(E_i)}{I^{(m^i_p,m^i_n)}(E_i)} \left\langle[\mathscr{O}^{DGT}(2)]^\dagger\mathscr{O}^{DGT}(2)\right\rangle^{(\widetilde{m_p},\widetilde{m_n})_i}\,.
\end{aligned} \tag{22}$$

It is easy to apply Eq. (22) by using the trace equation (21) with the replacement $V^{J_0}_{--}(0\nu) \longrightarrow V^{J_0}_{--}(DGT)$. The first correction to Eq. (22) follows from E^i_c in Eq. (19) and the next higher one from σ^2_i, with the change $\mathscr{O}(2:0\nu) \longrightarrow \mathscr{O}^{DGT}(2)$. Refs. [11, 33, 35] give details of SSM for expectation values and it is well known that the first correction is usually essential. Therefore for a realistic application of SSM to double GT strength sums, again we need the traces in Eq. (19). However, in this application, it is possible to make progress if we can derive the propagation formula just for $\langle [\mathscr{O}(2:0\nu)]^{\dagger}\mathscr{O}(2:0\nu)H\rangle^{(\widetilde{m_p},\widetilde{m_n})_i}$ and this is being attempted. Note that for the simple situation with m nucleons in N sp states and given irreducible two-body operators $\mathbf{U}$ and $\mathbf{V}$ with $\mathbf{U}^{\dagger} \neq \mathbf{U}$ and $\mathbf{V}^{\dagger} = \mathbf{V}$, the propagation equation for $\langle \mathbf{U}^{\dagger}\mathbf{U}\mathbf{V}\rangle^{m}$ follows easily by recognizing that the trace is a sixth order polynomial in m and $\langle \mathbf{U}^{\dagger}\mathbf{U}\mathbf{V}\rangle^{m'} = 0$ for $m' = 0$, 1, N and $N-1$. In terms of the symbol $(X)_r = (X)(X-1)\ldots(X-r+1)$, the final propagation formula is $\langle \mathbf{U}^{\dagger}\mathbf{U}\mathbf{V}\rangle^{m} = [(N)_5]^{-1}[(m)_2(N-m)_3 - (m)_3(N-m)_3]\langle\langle \mathbf{U}^{\dagger}\mathbf{U}\mathbf{V}\rangle\rangle^{2} + [(N)_6]^{-1}[(m)_4(N-m)_2]\langle\langle \mathbf{U}^{\dagger}\mathbf{U}\mathbf{V}\rangle\rangle^{N-2} + [(N)_6]^{-1}[(m)_3(N-m)_3]\langle\langle \mathbf{U}^{\dagger}\mathbf{U}\mathbf{V}\rangle\rangle^{3}$.

6. Conclusions

In this paper first given is an overview of nuclear models for NSME calculations for NDBD, with particular reference to shell model and group theoretical models. More importantly statistical spectroscopy method for calculating NSME for NDBD and also for the related double GT strength sums is introduced and the basic formalism is given in detail. To proceed further with SSM, propagation formulas for the traces in Eqs. (19, 20) are needed and also it is desirable to have a formalism to incorporate J-decomposition in Eq. (16). These problems are being addressed and we will report in the near future the full SSM theory for NDBD and double GT strength sums. Finally, an advantage of SSM is that it applies easily to transitions from ground 0^+ to excited states (also excited to excited states) of the daughter nucleus in DBD such as those in the recent ^{100}Mo experiment [36].

Acknowledgment

Thanks are due to K. Kar and R.U. Haq for many useful discussions.

References

[1] M. Goeppert-Mayer, Phys. Rev. **48**, 512 (1935).

[2] E. Majorana, Nuovo Cim. **14**, 171 (1937).

[3] G. Racah, Nuovo Cim. **14**, 322 (1937).

[4] W.H. Furry, Phys. Rev. **56**, 1184 (1939).

[5] H.V. Klapdor-Kleingrothaus, I.V. Krivosheina, A. Dietz, and O. Chkvorets, Phys. Lett. **B586**, 198 (2004).

[6] C. Arnaboldi et al., Phys. Rev. Lett. **95**, 142501 (2005).

[7] J. Suhonen and O. Civitarese, Phys. Rep. **300**, 1123 (1998).

[8] S.R. Elliott and P. Vogel, Ann. Rev. Nucl. Part. Phys. **52**, 115 (2002).

[9] S.R. Elliott and J. Engel, J. Phys. **G30**, R183 (2004).

[10] V.I. Tretyek and Y.G. Zdessenko, Atomic Data and Nuclear Data Tables **80**, 83 (2002).

[11] V.K.B. Kota, Phys. Rep. **347**, 223 (2001); Ann. Phys. (N.Y.) **306**, 58 (2003).

[12] J.B. French and V.K.B. Kota, Phys. Rev. Lett. **51**, 2183 (1983); J.B. French, S. Rab, J.F. Smith, R.U. Haq, and V.K.B. Kota, Can. J. Phys. **84**, 677 (2006).

[13] J.B. French, V.K.B. Kota, A. Pandey, and S. Tomsovic, Ann. Phys. (N.Y.) **181**, 235 (1988); V.K.B. Kota and K. Kar, Pramana-J. Phys. **32**, 647 (1989); V.K.B. Kota and D. Majumdar, Z. Phys. **A351**, 377 (1995).

[14] J. Engel, P. Vogel, and M.R. Zirnbauer, Phys. Rev. C **37**, 731 (1988).

[15] V.K.B. Kota and D. Majumdar, Z. Phys. **A351**, 365 (1995).

[16] O. Civitarese and J. Suhonen, Phys. Rev. C **47**, 2410 (1993).

[17] O. Castaños, J.G. Hirsch, O. Civitarese, and P.O. Hess, Nucl. Phys. **A571**, 276 (1994).

[18] O. Civitarese and J. Suhonen, Nucl. Phys. **A761**, 313 (2005).

[19] Y. Sun, Int. J. Mod. Phys. **E15**, 1695 (2006).

[20] E. Caurier, to appear in Prog. Part. Nucl. Phys. (2007); E. Caurier, F. Nowacki, and A. Poves, Int. J. Mod. Phys. **E16**, 552 (2007); Caurier, F. Nowacki, A. Poves, and J. Retamosa, Phys. Rev. Lett. **77**, 1954 (1996).

[21] M. Dufour and A.P. Zuker, Phys. Rev. C **54**, 1641 (1996).

[22] D.J. Dean, T. Engeland, M. Hjorth-Jensen, M.P. Kartamyshev, and E. Osnes, Prog. Part. Nucl. Phys. **53**, 419 (2004).

[23] S.E. Koonin, D.J. Dean, and K. Langanke, Phys. Rep. **278**, 1 (1997).

[24] I. Talmi, *Simple Models of Complex Nuclei - The Shell Model and the Interacting Boson Model* (Harwood, New York, 1993).

[25] K.T. Hecht, Nucl. Phys. **63**, 177 (1965); Phys. Rev. B **139**, 794 (1965); Nucl. Phys. **A102**, 11 (1967).

[26] J. Engel and P. Vogel, Phys. Rev. C **69**, 034304 (2004).

[27] V.K.B. Kota and J.A. Castilho Alcarás, Nucl. Phys. **A764**, 181 (2006); D.J. Rowe and M.J. Carvalho, J. Phys. **A40**, 471 (2007).

[28] J. Engel, S. Pittel, M. Stoitsov, P. Vogel, and J. Dukelsky, Phys. Rev. C **55**, 1781 (1997).

[29] O. Scholten and Z.R. Yu, Phys. Lett. **B161**, 13 (1985).

[30] J.G. Hirsch, Rev. Mex. Fis. **48**, 87 (2002).

[31] V.K.B. Kota and U. Datta Pramanik, Euro. Phys. Jour. **A3**, 243 (1998); R. Bijker and V.K.B. Kota, Ann. Phys. (N.Y.) **187**, 148 (1988).

[32] V.K.B. Kota and D. Majumdar, Nucl. Phys. **A604**, 129 (1996).

[33] S.S.M. Wong, *Nuclear Statistical Spectroscopy* (Oxford University Press, New York, 1986); V. Potbhare, Ph.D. Thesis (University of Rochester, 1975), unpublished.

[34] P. Vogel, M. Ericson, and J.D. Vergados, Phys. Lett. **B212**, 259 (1988); K. Muto, Phys. Lett. **B277**, 13 (1992).

[35] J.P. Draayer, J.B. French, and S.S.M. Wong, Ann. Phys. (N.Y.) **106**, 472 (1977).

[36] M.J. Hornish, L. De Braeckeleer, A.S. Barabash, and V.I. Umatov, Phys. Rev. C **74**, 044314 (2006).

NEUTRINOLESS DOUBLE BETA DECAY
Editors: V.K.B. Kota and U. Sarkar

STUDY OF DOUBLE BETA DECAY IN MEAN FIELD APPROACH - PRESENT AND FUTURE PERSPECTIVE

P. K. Rath

Department of Physics, University of Lucknow, Lucknow - 226007, India

Abstract The nuclear models are generically mean field models. However, the models in which the mean field is generated self consistently are Hartree-Fock, Deformed Hartree-Fock, Projected Hartree-Fock-Bogoliubov as well as QRPA and its extensions. The approximations involved in different mean field models are critically examined. The $\beta^-\beta^-$ decay of ^{94,96}Zr, ^{100}Mo, ^{110}Pd, ^{128,130}Te and ^{150}Nd isotopes is studied in the PHFB model and constraints on $\langle m_\nu \rangle$, $\langle \lambda \rangle$ and $\langle \eta \rangle$ are extracted from the observed limits on half-lives of $0\nu\ \beta^-\beta^-$ decay.

1. Introduction

The story of neutrinos is rather chequered and quite exciting due to their enigmatic nature. In the paradigm of "Standard Model" (SM), the neutrinos are assumed to be Weyl particles. Over the past years, a lot of information has been obtained on the mass and nature of neutrinos through terrestrial and extraterrestrial experiments. The mass of electron neutrino $m_{\nu_e} < 2.2\ eV$ [1], muon neutrino $m_{\nu_\mu} < 0.19\ MeV$ [2] and tao neutrino $m_{\nu_\tau} < 18.2\ MeV$ [3]. The analysis of WMAP and 2dFGRS data imply a constraint on the sum of neutrino masses $\sum_i m_i < 0.7\ eV$ [4]. Similarly, the obtained bound on $\sum_i m_i < 1.7\ eV$ from the WMAP and SDSS data [5]. The neutrino oscillation experiments provide us with neutrino mass square differences Δm^2 and mixing angles. The global analysis of solar, atmospheric and reactor neutrino data with 3σ error provides $\Delta m^2_{12} \simeq 7.9\times10^{-5}\ eV^2$, $\theta_{12} = 33.21°$ for solar neutrinos, $\Delta m^2_{23} \simeq 2.12^{+1.09}_{-0.81} \times10^{-3}$ eV^2, $\theta_{23} \simeq 45^{+10.55°}_{-9.33}$ for atmospheric neutrinos and $sin^2\theta_{13} \leq 0.022 \pm 0.047 \times 10^{-2}$ [6]. Specifically, the confirmation of flavor oscillation of neutrinos at atmospheric, solar, accelerator and reactor experiments suggest that the neutrinos are massive. Further, the oscillation data suggest that the neutrinos may belong to either normal hierarchy ($m_1 < m_2 < m_3$) or inverted hierarchy ($m_3 < m_1 < m_2$). The data do not exclude the possibility that the mass of the light neutrino is much larger than $\sqrt{\Delta m_a^2}$ and this implies the possible existence of quasi-degenerate neutrino mass spectrum i.e. $m_1 \simeq m_2 \simeq m_3$.

The massive neutrinos may be of Dirac or Majorana character. There are processes, which explicitly depend on the Majorana nature of neutrinos. In the other, it is however possible to extract information on the Majorana nature of neutrinos although the processes occur independent of the Dirac or Majorana character. The former processes are nuclear double beta ($\beta\beta$) decay, neutrinoless decay between a pair of single $\beta-$emitters given by ${}^A_ZX \rightarrow {}^A_{Z+1}Y + e^- + \overline{\nu}_e$ and subsequently, $\nu_e + {}^A_{Z+1}X \rightarrow {}^A_{Z+2}Y + e^-$ and $e^- + e^- \rightarrow W^- + W^-$. The lepton number conserving processes, which give information regarding the Majorana nature of neutrinos are (i) $\mu^- + {}^A_ZX \rightarrow {}^A_{Z-2}Y + e^+$, (ii) $\mu^+ \rightarrow e^+ + N_j + \overline{N}_j$, (iii) $\mu^+ + e^- \rightarrow \mu^- + e^+$ and (iv) $\nu_e - e^-$ scattering. The lepton number violating processes providing information on the Majorana nature of neutrinos are (i) $K^- \rightarrow \pi^+ + e^- + e^-$, (ii) $K^- \rightarrow \pi^+ + e^- + \mu^-$ and (iii) $K^- \rightarrow \pi^+ + \mu^- + \mu^-$. The magnetic dipole moment of neutrinos can also give information regarding the nature of neutrinos. However, it is generally agreed that the study of tritium single β-decay and nuclear $\beta\beta$ decay together can provide sharpest limits on the mass and nature of neutrinos. Specifically, the nuclear $\beta\beta$ decay can clarify a number of issues regarding the nature of neutrinos, namely the origin of neutrino mass (Dirac vs. Majorana), absolute scale on neutrino mass, the type of hierarchy and CP violation in the lepton sector.

2. Nuclear Double Beta Decay

The nuclear $\beta\beta$ decay is possible through the emission of electrons as well as positrons. In case of the latter, there are three possible kinematically competing modes, namely double-positron ($\beta^+\beta^+$) emission, electron-positron conversion (β^+EC) and double-electron capture ($ECEC$). The four experimentally distinguishable processes of $\beta\beta$ decay through their spectral shapes are two neutrino double beta ($2\nu\ \beta\beta$) decay, neutrinoless double beta ($0\nu\ \beta\beta$) decay, single Majoron accompanied neutrinoless double beta ($0\nu\ \beta\beta\phi$) decay and double Majoron accompanied neutrinoless double beta ($0\nu\ \beta\beta\phi\phi$) decay. The study of these four $\beta\beta$ decay modes have implications in nuclear structure physics, physics beyond the SM, the stellar evolution, history of the early universe, galaxy formation, big bang nucleosynthesis, thermal history of the universe, dark matter and dark energy. Presently, we deal with the former two issues by restricting the discussion to $\beta^-\beta^-$ decay modes only.

2.1. 2ν $\beta\beta$ Decay and Validity of Nuclear Models

The 2ν $\beta\beta$ decay conserves the lepton number exactly and it is allowed in the SM. The inverse half-life of the 2ν $\beta\beta$ decay for the $0^+ \rightarrow 0^+$ transition is given by

$$\left[T_{1/2}^{2\nu}(0^+ \rightarrow 0^+)\right]^{-1} = G_{2\nu}\left|M_{2\nu}\right|^2 \tag{1}$$

where the phase space factor $G_{2\nu}$ can be calculated with good accuracy and the nuclear transition matrix element (NTME) $M_{2\nu}$ defined as

$$M_{2\nu} = \sum_N \frac{\langle 0^+||\sigma\tau^\pm||1_N^+\rangle\langle 1_N^+||\sigma\tau^\pm||0^+\rangle}{E_N - (E_I + E_F)/2} \tag{2}$$

is nuclear model dependent [7].

The half-life $T_{1/2}^{2\nu}$ of 2ν $\beta^-\beta^-$ decay has been already measured for about 10 nuclei out of 35 possible candidates [8]. The values of $M_{2\nu}$ can be extracted directly from the observed half-lives of 2ν $\beta^-\beta^-$ decay. In all cases, it is observed that, the NTMEs are sufficiently quenched. The $M_{2\nu}$ extracted from the experimental data is approximately 0.005 to 0.115. The NTMEs $M_{2\nu}$ exhaust a small fraction (10^{-5} to 10^{-8}) of the double Gamow-Teller (DGT) sum rule and smaller than the single particle values by 1 to 3 orders of magnitude. The reduction in NTMEs by 10^{-1} to 10^{-3} from single particle values is believed to be due to the possible absorption of most of the 2ν $\beta\beta$ strength in GT resonance. Hence, it is sensitive to the details of the nuclear structure and the validity of different models employed for nuclear structure calculations can be tested by calculating the $M_{2\nu}$.

2.2. 0ν $\beta\beta$ Decay and Physics Beyond the SM

The possible occurrence of 0ν $\beta\beta$ decay requires two necessary conditions namely, finite neutrino mass and helicity matching so that the neutrino is a Majorana particle i.e. $\nu_{eL} = \overline{\nu}_{eL}$. The requirement of non-zero neutrino mass is contained quite naturally in theories beyond the *SM* and the Majorana mass term requires the breaking of lepton number *L*. Alternatively, the helicity matching is satisfied in the presence of right handed $V+A$ currents. The 0ν $\beta\beta$ decay is possible in gauge theoretical models beyond the SM, namely GUTs -Left-right symmetric models and E(6)-, R-parity conserving as well as violating Minimal supersymmetric models, in the scenarios of leptoquarks, compositeness and sterile neutrinos. The violation of Lorentz invariance (*VLI*) and the violation of weak equivalence principle (*VEP*) can be even studied through the 0ν $\beta\beta$ decay. The 0ν $\beta\beta\phi$ as well as 0ν $\beta\beta\phi\phi$ decay modes are possible in nine gauge theoretical models given by Bamert and coworkers [9].

In case of light neutrinos ($m_i < 10\,MeV$), the inverse half-lives of 0ν $\beta\beta$ decay for the $0^+ \rightarrow 0^+$ is given by

$$\left[T_{1/2}^{0\nu}(0^+ \to 0^+)\right]^{-1} = \left(\frac{\langle m_\nu \rangle}{m_e}\right)^2 C_{mm} + \left(\frac{< m_\nu >}{m_e}\right) < \lambda > C_{m\lambda}$$
$$+ \left(\frac{< m_\nu >}{m_e}\right) < \eta > C_{m\eta} + < \lambda >^2 C_{\lambda\lambda} + < \eta >^2 C_{\eta\eta}$$
$$+ < \lambda >< \eta > C_{\lambda\eta} \quad (3)$$

where the nuclear factors-of-merits C_{xy} are expressed as combinations of 9 NTMEs and 9 phase space factors G_{0k} [7]. One can extract limits on gauge theoretical parameters, namely effective neutrino mass $< m_\nu >$, effective parameters of right handed currents $< \lambda >$ and $< \eta >$ from the observed half-life of 0ν $\beta\beta$ decay. The 0ν $\beta^-\beta^-$ decay has not been observed so far and the best observed limits on half-lives $T_{1/2}^{0\nu} > 1.9 \times 10^{25}$ yr for ^{76}Ge [10]. However, Klapdor and his group have recently reported that 0ν $\beta^-\beta^-$ decay has been observed in ^{76}Ge and $T_{1/2}^{0\nu} = 1.19 \times 10^{25}$ yr [11]. The later result is controversial, but it is expected that the issue will be settled soon [12]. The aim of all the present experimental activities is to observe the 0ν $\beta^-\beta^-$ decay.

3. Nuclear Models

The NTMEs are usually calculated in the two-nucleon (2n) mechanism. In case, mesonic and isobaric degrees of freedom are allowed in addition, several other modes, namely $N \longleftrightarrow \Delta$ mechanism, $\Delta \longleftrightarrow \Delta$ mechanism and $\pi^- \longleftrightarrow \pi^+$ mechanism have to be considered. The contribution of the latter will be important if the NTMEs are small in 2n mechanism. Over the past few years, several nuclear models have been employed to calculate the $\beta\beta$ decay rates in 2n mechanism. These models can be classified into three broad categories, namely shell model and its variants, mean field models and alternative models. In the below, we discuss the relative applicability, success and failure of various models used so far to study the $\beta\beta$ decay modes.

3.1. Shell Model and Its Variants

The shell model attempts to solve the nuclear many-body problem as exactly as possible, and is the best choice for the calculation of the NTMEs for $\beta\beta$ decay modes. In the shell model, the number of basis states beyond the *pf*-shell increases so drastically that a few years back it was not possible to perform a conventional shell model calculation without imposing certain truncation scheme. On the other hand, most of the $\beta\beta$ decay emitters are medium or heavy mass nuclei. Hence, a reliable shell-model calculation was difficult to perform. The presently used large scale shell model calculations are more promising in nature [13].

In shell model calculations, the examination the model space dependence of NTMEs $M_{2\nu}$ in ^{48}Ca [14] and ^{92}Mo [15] show that stringent truncation imposed on number of shell model basis states prevent configurations responsible for the reduction of NTMEs by destructive interference. Further, the omission of spin-orbit partners may also hinder the destructive interference between spin-flip and spin non-flip contributions leading to the suppression of NTMEs for 2ν $\beta^-\beta^-$ decay. In case of 0ν

$\beta^-\beta^-$ decay, the shell model calculations show in fact that the closure approximation is valid. Further, all multipole up to large angular momenta up to $5\hbar$ are involved and have comparable contribution.

The shell model calculations, besides being feasible for a limited number of nuclei due to a large number of basis states, fail to fulfil Ikeda sum rule by about 40 to 60 %. The renormalization of vector and axial vector weak interaction strengths g_V and g_A respectively due to nuclear medium effects results in $g_V/g_A \approx 1$. This leads to renormalization of Gamow-Teller operator, which is responsible for quenching of about 60%. Thus, the mechanism responsible for the quenching of $2\nu\ \beta^-\beta^-$ decay strength is not well understood.

3.2. Mean Field Models

The nuclear models are generically mean field models. However, one takes a phenomenological mean field in the shell model for convenience. On the other hand, the mean field is generated through self consistency in HF, DHF, PHFB as well as QRPA and its extensions. Using Wick's theorem, the many body Hamiltonian can be written as

$$H = H_0 + H_2 + H_4 \tag{4}$$

where

$$H_0 = \sum_{\alpha\beta} \langle\alpha| \left(T + \frac{1}{2}\mathscr{V}\right) |\beta\rangle \langle a^\dagger_\alpha a_\beta \rangle \tag{5}$$

$$\begin{aligned} H_2 &= \sum_{\alpha\beta} \langle\alpha| (T + \mathscr{V}) |\beta\rangle : a^\dagger_\alpha a_\beta : + \frac{1}{4} \sum_{\alpha\beta\gamma\delta} \langle\alpha\beta|V|\gamma\delta\rangle \left\langle a^\dagger_\beta a_\delta \right\rangle : a^\dagger_\alpha a_\gamma : \\ &+ \frac{1}{4} \sum_{\alpha\beta\gamma\delta} \left[\langle\alpha\beta|V|\gamma\delta\rangle \langle a_\delta a_\gamma\rangle : a^\dagger_\alpha a^\dagger_\beta : + h.c. \right] \end{aligned} \tag{6}$$

$$H_4 = \frac{1}{4} \sum_{\alpha\beta\gamma\delta} \langle\alpha\beta|V|\gamma\delta\rangle : a^\dagger_\alpha a^\dagger_\beta a_\delta a_\gamma : \tag{7}$$

DHF Model In the HF and DHF approach, one solves Eq. (5) only, neglecting H_2 and H_4. In the DHF model, the deformed intrinsic state $|\ \phi_K\rangle$ is a Slater determinant for protons and neutrons consisting of deformed orbits $|\eta\ m\rangle$ given by

$$|\ \eta m\rangle = \sum_j C_{jm} \mid jm\rangle \tag{8}$$

where the amplitude C_{jm} of the spherical state $|\ jm\rangle$ in the orbit η is obtained by solving the deformed Hartree-Fock equations. Each deformed intrinsic state is a superposition of states of good angular momenta, which is obtained by projection technique. The states $\left|\psi_K^{J_1M}\right\rangle$ and $\left|\psi_K^{J_2M}\right\rangle$ with definite angular momenta J_1 and J_2 projected form various intrinsic states $|\phi_{K_1}\rangle$and $|\phi_{K_2}\rangle$ are in general not orthogonal to each other even although they are orthogonal before angular momentum projection. However, if the Hamiltonian is diagonalized in the subspace generated from the projected states then the resulting eigenstates are orthogonal, and the matrix elements of various tensor operators can be calculated with these orthonormalized states .

In the DHF, Sahu *et al.* have studied the $2\nu\ \beta^-\beta^-$ decay of ^{76}Ge [16]. The effects of deformation on the distribution of the Gamow-Teller (GT) and β-decay properties has been studied using a quasiparticle Tamm-Dancoff approximation (TDA) based on DHF calculations with Skyrme interactions [17].

PHFB Model The essential idea behind the HFB theory is to transform particle coordinates to quasiparticle coordinates through general Bogoliubov transformation such that the quasiparticles are relatively weakly interacting. Essentially, the Hamiltonian H is expressed as

$$H = H_0 + H_{qp} + H_{qp-qp} \tag{9}$$

where H_0 is the energy of the quasiparticle vacuum, H_{qp} is the elementary quasiparticle excitations and H_{qp-qp} is the interaction between quasiparticles, which is weak. In HFB theory, the interaction between the quasiparticles is usually neglected and the Hamiltonian H is approximated by an independent quasiparticle Hamiltonian. In TD-HFB or QRPA, some effects of quasiparticle interaction can be included.

Tomoda *et al.* have studied the $2\nu\ \beta^-\beta^-$ and $0\nu\ \beta^-\beta^-$ decay of ^{76}Ge in the PHFB model using MSDI in the closure approximation [18]. The $M_{2\nu}$ was not sufficiently quenched. The closure is a valid approximation for the $0\nu\ \beta^-\beta^-$ decay, but the NTMEs are larger than the corresponding QRPA values. However, further studies in this model were not pursued. Dixit *et al.* have successfully calculated the $2\nu\ \beta^-\beta^-$ decay rate of ^{100}Mo in the PHFB model, and it has been shown that the deformations of the intrinsic ground states of ^{100}Mo and ^{100}Ru play a crucial role in reproducing a realistic $M_{2\nu}$ of ^{100}Mo [19]. The $0^+ \rightarrow 2^+$ transition of $2\nu\ \beta^-\beta^-$ decay has also been studied in the same PHFB model [20, 21]. Further, the PHFB model has been applied, in conjunction with the summation method, to study the $2\nu\ \beta^-\beta^-$ decay for the $0^+ \rightarrow 0^+$ transition in the mass region $A \sim 100$ [22] and $A \sim 150$ [23]. It has been shown that the deformations of the intrinsic ground states play a crucial role in reproducing a realistic $M_{2\nu}$. It has been also shown that the deformation effects are important in the case of $0\nu\ \beta^-\beta^-$ decay [24].

QRPA and Its Extensions Vogel and Zirnbauer were the first to provide an understanding of the observed suppression of $M_{2\nu}$ in the QRPA model [25]. In $\beta^-\beta^-$ decay, the initial nucleus decays into the final nucleus through the virtual excitation of all possible states of the intermediate odd-odd nucleus. The excitation energies of even-even nuclei are obtained taking two quasiparticle excitations built of quasi-proton or quasi-neutron pairs into account. The states of adjacent odd-odd nuclei are generated by use of charge changing mode of the QRPA i.e. the proton-neutron quasiparticle excitations (*pn*-QRPA) of the adjacent even-even ground state. More explicitly, the proton-neutron particle-hole (*p-h*) or proton-neutron particle-particle (*p-p*) interaction matrix elements are required to calculate the excited states of the intermediate nucleus.

The repulsive *p-h* and attractive *p-p* interactions are responsible for the concentration of the β^- strength in the giant *GT* resonance and the reduction of total β^+ strength and its concentration at low excitation energies respectively. The *p-p* interaction has negligible effect on the strength of the giant *GT* resonance and was usually neglected. It was observed that the quenching of $M_{2\nu}$ can be achieved by a proper inclusion of ground state correlations through the *p-p* interaction in the $S = 1$, $T = 0$ channel and the calculated half- lives are in close agreement with all the experimental data.

In the QRPA, a large number of intermediate states can be constructed in a small model space taking advantage of the harmonic approximation. However, any small change in phenomenological parameter g_{pp} causes a large change in lifetimes. The entire method collapses, when the value of this parameter exceeds a critical value and the physical value of this parameter is usually close to the point at which the QRPA solutions collapse. The origin of this unfortunate behavior is attributed to the quasi-boson approximation (QBA) of the QRPA operators violating the Pauli exclusion principle. The most important proposals directed towards solving some of the inherent problems are (i) Full QRPA (FQROPA) - with the inclusion of proton-neutron pairing, (ii) RQRPA and FRQRPA - in which violation of Pauli exclusion principle is restored, (iii) Higher QRPA (HQRPA), (iv) Multiple commutator method (MCM) and (v) Particle number projection (PQRPA) [26].

In the case of 0ν $\beta^-\beta^-$ decay, the QRPA calculations show the mentioned features of the shell model. In addition, the QRPA calculations exhibit two additional features, namely (i) the NTMEs are less sensitive to the particle-particle coupling constant g_{pp} than the 2ν $\beta^-\beta^-$ decay, as the contribution of the 1^+ multipole is relatively small and (ii) the short ranged nucleon-nucleon repulsion has to be included in spite of the fact that the neutrino potential is long ranged.

3.3. Alternative Models

The shell model and its variants as well as the QRPA and its extensions, besides having their own limitations, fail to fulfil the Ikeda sum rule. In the QRPA approach, a larger uncertainty may occur in deformed nuclei as deformation is usually ignored. This has been shown for ^{100}Mo and ^{150}Nd using the QRPA [27]. The small predictive power of the QRPA and its extensions is the main motivation for seeking an alternative description which might result in a more reliable approach for the study of nuclear $\beta\beta$ decay processes. These attempts have been summarized as alternative models, which are operator expansion method (OEM), two-vacua RPA (TVRPA), group theoretical methods, namely SU(2), SU(4)$_{\sigma\tau}$, SO(5), O(8) and pseudo-SU(3) and single state dominance hypothesis (SSDH). Presently, the models in wide use are SU(4)$_{\sigma\tau}$, pseudo-SU(3) and SSDH.

4. Results and Discussions

The present aim of all theoretical calculations is to evaluate reliable NTMEs for the 0ν $\beta^-\beta^-$ decay so that various gauge-theoretical parameters, namely $< m_\nu >$, $\langle\lambda\rangle$ and $\langle\eta\rangle$ etc. can be extracted as accurately as possible from the observed limits on half-lives $T^{0\nu}_{1/2}$ of 0ν $\beta^-\beta^-$ decay. There is no objective way to judge a priori the correctness of theoretical calculations. The common practice is to calculate the $M_{2\nu}$ at first and compare with the experimentally observed value, because the 2ν $\beta^-\beta^-$ and 0ν $\beta^-\beta^-$ decay modes involve the same set of initial and final nuclear wave functions.

Table 1. Experimental and theoretical half-lives $T^{2\nu}_{1/2}$ (yr) of ^{94,96}Zr, ^{100}Mo, ^{110}Pd, ^{128,130}Te and ^{150}Nd nuclei for the $0^+ \to 0^+$transition. (a) and (b) correspond to $T^{2\nu}_{1/2}$ calculated for $g_A = 1.25$ and 1.0 respectively.

	Experiment			Theory		
Nuclei	Project	$T^{2\nu}_{1/2}$		$\|M_{2\nu}\|$		$T^{2\nu}_{1/2}$
^{94}Zr	NEMO 2	$>1.1\times10^{-5}$	[28]	0.076	(a)	7.51
(10^{22} yr)					(b)	18.34
^{96}Zr	NEMO 3	$2.0\pm0.3\pm0.2$	[29]	0.058	(a)	1.56
(10^{19} yr)	Avearage	2.0 ± 0.3	[30]		(b)	3.80
^{100}Mo	NEMO 3	$7.72\pm0.02\pm0.54$	[29]	0.104	(a)	9.79
(10^{18} yr)	ITEP+INFN	$7.2\pm0.9\pm1.8$	[31]		(b)	23.90
	UC Irvine	$6.82^{+0.38}_{-0.53}\pm0.68$	[32]			
	LBL Collab.	$7.6^{+2.2}_{-1.4}$	[33]			
	ELEGANTS V	$11.5^{+3.0}_{-2.0}$	[34]			
	Average	7.1 ± 0.4	[30]			
^{110}Pd		$>6.0\times10^{-4}$	[35]	0.133	(a)	1.41
(10^{20} yr)					(b)	3.44
^{128}Te	Recommended	2.5 ± 0.3	[30]	0.033	(a)	1.05
(10^{24} yr)					(b)	2.55
^{130}Te	Milano+INFN	$6.1\pm1.4^{+2.9}_{-3.5}$	[36]	0.042	(a)	1.16
(10^{20} yr)	Recommended	9.0 ± 1.0	[30]		(b)	2.82
^{150}Nd	NEMO 3	$9.7\pm0.7\pm1.0$	[29]	0.0245	(a)	14.1
(10^{18} yr)	UCI	$6.75^{+0.37}_{-0.42}\pm0.68$	[32]		(b)	34.3
	ITEP +INR	$18.8^{+6.6}_{-3.9}\pm1.9$	[37]			
	Averag	7.8 ± 0.7	[30]			

However, the structure of operators in the two decay modes are completely different. While the NTME of 2ν $\beta^-\beta^-$ decay is spin-isospin dependent, the NTMEs of $\beta^-\beta^-$ decay are energy-momentum dependent as well. Hence, the reliability of NTMEs can be judged a priori only on the basis of the success of a nuclear model in explaining as much observed physical properties of nuclei as possible in accordance with the basic philosophy of nuclear many-body theory.

We test the reliability of PHFB wave functions by comparing the calculated yrast spectra, reduced $B(E2:0^+ \to 2^+)$ transition probabilities , static quadrupole moments $Q(2^+)$ and the gyromagnetic factors $g(2^+)$ ^{94,96}Zr, ^{94,96,100}Mo, ^{100}Ru, ^{110}Pd, ^{110}Cd,

Table 2. Limits on $\langle m_\nu \rangle$, $\langle \lambda \rangle$ and $\langle \eta \rangle$ form the experimentally observed limits on half-lives $T_{1/2}^{0\nu}$ of the 0ν $\beta^-\beta^-$ decay for ^{94,96}Zr, ^{100}Mo, ^{110}Pd, ^{128,130}Te and ^{150}Nd nuclei. (a) and (b) correspond to values extracted using NTMEs for point nucleons and with finite size effect as well as short range correlations respectively

Nuclei	Best limit on $T_{1/2}^{0\nu}$(yr)	Ref.	.	$\langle m_\nu \rangle$ (eV)	$\langle \lambda \rangle$	$\langle \eta \rangle$
^{94}Zr	$>1.9\times10^{19}$	[28]	(a)	1438.7	4.3×10^{-3}	2.3×10^{-5}
			(b)	1512.7	8.7×10^{-4}	4.7×10^{-6}
^{96}Zr	$>1.0\times10^{21}$	[28]	(a)	46.23	2.4×10^{-7}	2.2×10^{-7}
			(b)	44.64	4.9×10^{-8}	4.4×10^{-8}
^{100}Mo	$>4.6\times10^{23}$	[38]	(a)	1.09	1.2×10^{-6}	1.9×10^{-8}
			(b)	1.10	2.4×10^{-7}	3.9×10^{-9}
^{110}Pd	$>6.0\times10^{17}$	[35]	(a)	1464.9	2.5×10^{-9}	2.5×10^{-5}
			(b)	1500.1	4.9×10^{-9}	5.0×10^{-6}
^{128}Te	$>1.1\times10^{23}$	[36]	(a)	22.4	9.2×10^{-9}	4.1×10^{-7}
			(b)	23.9	1.8×10^{-9}	8.2×10^{-8}
^{130}Te	$>1.8\times10^{24}$	[39]	(a)	0.82	1.1×10^{-6}	1.7×10^{-8}
			(b)	0.83	2.2×10^{-7}	3.4×10^{-9}
^{150}Nd	$>1.22\times10^{21}$	[32]	(a)	26.78	2.7×10^{-5}	6.6×10^{-7}
			(b)	26.73	5.5×10^{-6}	1.3×10^{-7}

^{128,130}Te, ^{128,130}Xe, ^{150}Nd and ^{150}Sm participating in the 0ν $\beta^-\beta^-$ decay, with the available experimental data. The overall agreement between the theoretically calculated and observed spectroscopic properties of nuclei are reasonably good. Subsequently, the PHFB wave functions are employed to study the 2ν $\beta^-\beta^-$ decay for the $0^+ \rightarrow 0^+$ transition and compared with observed half-lives (Table 1) as a further test of reliability [22]. Finally, we extract limits on effective gauge theoretical parameters $< m_\nu >$, $\langle \lambda \rangle$ and $\langle \eta \rangle$, which are displayed in Table 2. The study of single electron energy spectrum and angular correlation will be helpful in distinguishing the role of m_ν, λ and η mechanism after the observation of the 0ν $\beta^-\beta^-$ decay.

The half-life of the 0ν $\beta^-\beta^-$ decay can be predicted using the observed parameters of neutrino oscillation and other methods aiming to measure the mass of the neutrino [40]. The predicted half-lives $T_{1/2}^{0\nu}$ of 0ν $\beta^-\beta^-$ decay for the three neutrino mass spectrum using the NTMEs calculated in different nuclear models show that the decay rates of 0ν $\beta^-\beta^-$ due to quasi-degenerate and inverted hierarchy neutrinos can be realized at ongoing and proposed experiments. However, the $T_{1/2}^{0\nu}$ of 0ν $\beta^-\beta^-$ decay due to normal hierarchy neutrinos can not be realized in near future. Taking

Klapdor's claim that the 0ν $\beta^-\beta^-$ decay of ^{76}Ge has been observed at face value [11], one gets an effective neutrino mass $\langle m_\nu \rangle = 0.397$ eV $\langle \lambda \rangle = 1.032\times10^{-7}$ and $\langle \eta \rangle = 6.600\times10^{-10}$ using world average NTMEs. The effective neutrino mass 0.397 eV corresponds to quasi degenerate neutrinos. The half-lives $T_{1/2}^{0\nu}$ of presently studied nuclei can be predicted using these extracted parameters. It is seen that the Klador's claim can be verified in the 0ν $\beta^-\beta^-$ decay experiments on ^{100}Mo and ^{130}Te under NEMO and CUORICINO projects respectively, although Klapdor's claim in ^{76}Ge nucleus is supposed to be verified by IGEX experiment.

5. Conclusions

It is observed that the validity of nuclear models presently employed to calculate the NTME $M_{2\nu}$ as well as half-life $T_{1/2}^{2\nu}$ can not be uniquely established due to large error bars in experimental results as well as uncertainty in g_A. The typical uncertainty in the experimental observation as well as theoretical calculations is of the order of 2-3. Hence, further work is necessary both in the experimental as well as theoretical front to judge the relative applicability, success and failure of various models used so far for the study of 2ν $\beta^-\beta^-$ decay processes. The 0ν $\beta^-\beta^-$ decay has been studied in the light neutrino approximation and stringent limits have been obtained on $< m_\nu >$, $\langle \lambda \rangle$ and $\langle \eta \rangle$ from the observed limits on 0ν $\beta^-\beta^-$ decay half-lives. The deformation effects are important both for the 2ν $\beta^-\beta^-$ and 0ν $\beta^-\beta^-$ decay. The accuracy of extracted gauge theoretical parameters depend on the reliability of NTMEs. Therefore, it is desirable that the uncertainty in NTMEs should be reduced as much as possible.

References

[1] C. Weinheimer, hep-ex/0210050; J. Bonn *et al.*, Prog. Part. Nucl. Phys. **48**, 133 (2002); V. M. Lobashev *et al.*, Phys. Lett. B **460**, 227 (1999); Nucl. Phys. Proc. Suppl. **91**, 280 (2001).

[2] D. E. Groom *et al.*, (Particle data group) Eur. Phys. J. C **15**, 1 (2000).

[3] R. Barate *et al.*, EPJ C **2**, 395 (1998).

[4] C. L. Bennet *et al.*, astro-ph/0302207; D. N. Spergel *et al.*, astro-ph/0302209.

[5] M. Tegmark *et al.*, astro-ph/0310723.

[6] M. Maltoni, T. Schwarz, M. A. Tortola and J. W. F. Valle, New J. Phys. **6**, 122 92004, hep-ph/0405172.

[7] M. Doi, T. Kotoni and E. Takasugi, Prog. Theor. Phys. Suppl. **83**, 1 (1985).

[8] V. I. Tretyak and Y. G. Zdesenko, Atomic data and Nuclear data Tables, **61**, 43 (1995), ibid **80**, 83 (2002).

[9] P. Bamert, C. P. Burgess and R. N. Mohapatra, Nucl. Phys. B **449**, 25 (1995).

[10] H. V. Klapdor-Kleingrothaus *et al.*, Mod. Phys. Lett. A **16**, 2409 (2001); hep-ph/0103062.

[11] H. V. Klapdor-Kleingrothaus, A. Dietz and I. V. Krivoshina, Phys. Rev. D **70**, 078301 (2004).

[12] C. E. Aalseth *et al.*, Phys. Rev. D **65**, 092007 (2002); hep-ph/0202018; R. Elliot and J. Engel, hep-ph/0005078.

[13] E. Caurier, A. Poves and A. P. Zuker, Phys. Lett. B **252**, 13 (1990); E. Caurier, F. Nowacki, A. Poves and J. Retamosa, Phys. Rev. Lett. **77**, 1954 (1996); E. Caurier, F. Nowacki, A. Poves and J. Retamosa, Nucl. Phys. A **654**, 973 (1999); L. Zhao, B. A. Brown and W. A. Richter, Phys. Rev. C **42**, 1120 (1990); L. Zhao and B. A. Brown, Phys. Rev. C **47**, 2641 (1993).

[14] L. Zhao, B. A. Brown and W.A. Richter, Phys. Rev. C **42**, 1120 (1990).

[15] J. Suhonen, P. C. Divari, L. D. Skouras and I. P. Johnstone, Phys. Rev. C **55**, 714 (1997).

[16] R. Sahu, F. Simkovic and A. Faessler, J. Phys. G. Nucl. Part Phys. **25**, 1159 (1999).

[17] F. Frisk, I. Hamamoto and X. Z. Zhang, Phys. Rev. C **52**,2468 (1995).

[18] T. Tomoda, A. Faessler, K. W. Schmid and F. Gruemmer, Nucl. Phys. A **452**, 591 (1986).

[19] B. M. Dixit, P. K. Rath and P. K. Raina, Phys. Rev. C **65**, 034311 (2002); ibid **67**, 059901 (E) (2003).

[20] K. Chaturvedi, B. M. Dixit, P. K. Rath and P. K. Raina, Phys. Rev. C **67**, 064317 (2003).

[21] J. Singh, R. Chandra, P. K. Raina and P. K. Rath , Pramana Journal of Physics **65**, 517 (2005).

[22] R. Chandra, J. Singh, P. K. Rath, P. K. Raina and J. G. Hirsch, Euro. Phys. J. A **23**, 223 (2005).

[23] S. Singh, R. Chandra, P. K. Rath, P. K. Raina and J. G. Hirsch (to be submitted).

[24] P. K. Rath, Proceedings of 1st Yamada Symposium on Neutrinos and Dark Matter in Nuclear Physics, Osaka University, Osaka, Japan (2003).

[25] P. Vogel and M. R. Zirnbauer, Phys. Rev. Lett. **57**, 3148 (1986).

[26] J. Suhonen and O. Civitarese, Phys. Rep. **300**, 123 (1998); A. Faessler and F. Simkovic, J. Phys. G **24,** 2139 (1998); hep-ph/9901215.

[27] A. Griffiths and P. Vogel, Phys. Rev. C **46**, 181 (1992); J. Suhonen and O. Civitarese, Phys. Rev. C **49**, 3055 (1994).

[28] R. Arnold *et al.*, Nucl. Phys. A **658**, 299 (1999).

[29] D. Lalanne, hep-ex/0509005.

[30] A. S. Barabash, nucl-ex/0602009.

[31] V. D. Ashitkov *et al.*, JETP Lett. **74**, 529 (2001).

[32] A. De Silva, M.K. Moe, M.A. Nelson and M.A. Vient, Phys. Rev. C **56**, 2451 (1997).

[33] M. Alston-Garnjost *et al.*, Phys. Rev. C **55**, 474 (1997).

[34] II. Ejiri *et al.*, J. Phys. G **17**, 155 (1991).

[35] R. G. Winter, Phys. Rev. **85**, 687 (1952).

[36] C. Arnaboldi *et al.*, Phys. Lett. B **557**, 167 (2003).

[37] V. Artemiev *et al.*, Phys. Lett. B **345**, 564 (1995).

[38] R. Arnold *et al.*, Phys. Rev. Lett. **95**, 182302 (2005).

[39] C. Arnaboldi *et al.*, Phys. Rev. Lett. **95**, 142501 (2005).

[40] S. M. Bilenky, A. Fassler and F. Simkovic, hep/ph-0402250.

NEUTRINOLESS DOUBLE BETA DECAY
Editors: V.K.B. Kota and U. Sarkar

DEFORMED SHELL MODEL RESULTS FOR DOUBLE BETA DECAY MATRIX ELEMENTS

R. Sahu[1]

Physics Department, Berhampur University, Berhampur - 760 007, Orissa, India

Abstract

The two neutrino double beta decay ($2\nu\beta\beta$-decay) for $^{76}Ge \rightarrow {}^{76}Se$ is studied within the framework of the deformed Shell model based on Hartree-Fock states. An interaction given by Kuo and modified by Ahalpara et al is used. The spectroscopic properties are well reproduced within the model. The calculated Gamow-Teller matrix elements for $0^+ \rightarrow 0^+$ and $0^+ \rightarrow 2^+$ transitions are compared with experiment. The agreement with experiment is quite satisfactory.

1. Introduction

The study of nuclear double beta decay has been a fascinating problem in recent years. The developments in the field are dicussed in recent review articles [1, 2, 3]. Double beta decay is a second order process of weak interaction predicted long back by Maria

[1] **E-mail:rankasahu@rediffmail.com**

Meyer in the year 1935. It is a rare process in which the charge number of a nucleus changes by two units. The $2\nu\beta\beta$-decay in which two electrons and two antineutrinos are emitted does not violate lepton number and is fully consistent with the standard model of electroweak theory. On the other hand the $0\nu\beta\beta$-decay violates the lepton number conservation and has been predicted in some theories beyond the standard model. The neutrinoless double beta decay which has not been observed yet, is more intresting as it provides a tool to distinguish whether the neutrino is a Majorana or a Dirac particle. In order to deduce the lepton number violating parameters of interest from the experimental lower limits on the half-life of $0\nu\beta\beta$-decay, the corresponding nuclear transition should be calculated correctly. Hence one should first test the goodness of the nuclear model by calculating the $2\nu\beta\beta$-decay which is already well established experimentally for a few isotopes. Thus two neutrino double beta decay provides a testing ground for different nuclear models.

Over the years, the DSM formalism has been quite successful in describing many spectroscopic properties of medium heavy nuclei.It would be quite interesting to study the $^{76}Ge \rightarrow ^{76}Se$ double beta decay.

2. Theory

The Gamow-Teller matrix element $M_{GT}^{2\nu}$ for the $2\nu\beta\beta$-decay is given by

$$M_{GT}^{2\nu}(J^+) = \sum_m \frac{< J_f^+||\sigma\tau^+||1_m^+ >< 1_m||\sigma\tau^+||0_i^+ >}{(E_m - (E_i + E_f)/2)^s} \tag{1}$$

where 0_i^+ is the ground state of the parent nucleus with ground state energy E_i. J_f^+ is either 0^+ or 2^+ state of the daughter nucleus. 1_m^+ are the 1^+ states of the intermediate odd-odd nucleus and E_m are the energies of these 1^+ states. Since we restrict here only to the $2\nu\beta\beta$-decay of ^{76}Ge, 0_i^+ corresponds to the lowest 0^+ state of ^{76}Ge, J_f^+ the lowest 0^+ or 2^+ state of ^{76}Se and 1_m^+ corresponds to 1^+ states of the intermediate nucleus ^{76}As. Here s=1 for $0^+ \rightarrow 0^+$ transition and s=3 for $0^+ \rightarrow 2^+$ transition. We briefly discuss below the DSM formalism which is used to evaluate the above matrix element.

The nuclear wave functions $|0_i^+>$ and $|J_f^+>$ for the initial and final nuclei are obtained by projecting out good angular momentum states from the low-lying axially symmetric prolate HF intrinsic states of the corresponding nucleus and then performing a band mixing calculation for each nucleus separately. The deformed single particle orbits $|i>$ obtained after solving the HF single particle equation self-consistently are related to the basis states $|jk>$ through the relation

$$|i> = \sum_{j_i} c_{j_i k_i} |j_i k_i> \tag{2}$$

Here j_i is the spherical single particle angular momentum and k_i is its projection along the symmetry axis. In the above equation, k_i is not summed over since we restrict ourselves to axial symmetry and hence k_i is a conserved quantity. The HF intrinsic states $\chi_K(\eta)$, which is an antisymmetrized product of the deformed single-particle orbits, do not have definite angular momenta. Good angular momentum states are projected out from each of these intrinsic states using the angular momentum projection operator

$$P^J_{MK} = \frac{2J+1}{8\pi^2} \int d\Omega D^{J^*}_{MK}(\Omega) R(\Omega) \tag{3}$$

Here Ω represents the Euler angles (α, β, γ), $R(\Omega)$ which is equal to $\exp(-i\alpha J_z)\exp(-i\beta J_y)\exp(-i\gamma J_z)$ represents the general rotation operator. The normalized states of good angular momentum projected from the intrinsic state $\chi_K(\eta)$ can be written in the form

$$\psi^J_{MK}(\eta) = \frac{2J+1}{8\pi^2\sqrt{N_{JK}}} \int d\Omega D^{J^*}_{MK}(\Omega) R(\Omega) |\chi_K(\eta)> \tag{4}$$

where N_{JK} is the normalization constant given by

$$N_{JK} = \frac{2J+1}{2} \int_0^\pi d\beta sin\beta d^J_{KK}(\beta) < \chi_K(\eta)|e^{-i\beta J_y}|\chi_K(\eta)> \tag{5}$$

It is easy to see that

$$< \chi_{K_1}|e^{-i\beta J_y}|\chi_{K_2} >= det\,[M_{ik}(\beta)] \tag{6}$$

where

$$M_{ik}(\beta) = \sum_j c_{jk} c_{jk'} d^j_{kk'}(\beta) \tag{7}$$

The good angular momentum states $\psi^J_{MK}(\eta)$ projected from different intrinsic states $\chi_K(\eta)$ are, in general, not orthogonal to each other. They are orthonormalized by performing a band mixing calculation. The orthonormalized nuclear wave function $\phi^J_M(\alpha)$ is then written as

$$\phi^J_M(\alpha) = \sum_{K,\eta} S^J_{K,\eta} \psi^J_{MK}(\eta) \tag{8}$$

The label α distinguishes different states with the same angular momentum. The calculations are performed taking the band mixed wave function $\phi^J_M(\alpha)$. Now the reduced matrix element appearing in Eq. (1) can be evaluated using the formula

$$\begin{aligned} < \psi^{J_1}_{K_1}(\mu)||\sigma\tau^+||\psi^{J_2}_{K_2}(\eta)> &= \frac{2J_2+1}{2}\sqrt{\frac{2J_1+1}{N_{J_1K_1}N_{J_2K2}}} \begin{bmatrix} J_2 & 1 & J_1 \\ K_2 & \nu & K_1 \end{bmatrix} \\ &\int_0^\pi d\beta sin\beta d^{J_1}_{K1,K2+\nu} \\ &< \chi_{K_1}(\mu)|e^{-i\beta J_y}\sigma_\nu\tau^+|\chi_{K_2}(\eta)> \end{aligned} \tag{9}$$

The quantity between the square bracket represents the Clebsch-Gordon coefficient. Since $\sigma\tau^+ = \sum_{\lambda=1}^n \sigma_1(\lambda)\tau_1^+(\lambda)$, the kernel reduces to the form

$$\begin{aligned} < \chi_{K_1}(\mu)|e^{-i\beta J_y}\sigma_\nu\tau^+|\chi_{K_2}(\nu)> &= \sum_{i,k=1}^n (-1)^{i+k} D^{n-1}_{i,k} \sum_{j_i,j_k} \frac{(-1)^{(3j_i-j_k-\nu)}}{(2j_k+1)^{1/2}} \\ &c^*_{j_im_i} c_{j_km_k} d^{j_i}_{m_i,m_k+\nu} \\ &\begin{bmatrix} j_i & 1 & j_k \\ m_k+\nu & -\nu & m_k \end{bmatrix} \\ &< n_il_ij_i\tau_i|\sigma_1\tau_1^+|n_kl_kj_k\tau_k> \end{aligned} \tag{10}$$

Here $D^{n-1}_{i,k}$ is the determinant of rank $(n-1)$ of the matrix elements given by Eq. (7) obtained from the $n \times n$ determinant given in Eq. (6) by removing the ith row and kth column. It should be remembered that χ_{K_1} and χ_{K_2} correspond to HF intrinsic states of two different nuclei. The matrix element

$$< n_i l_i j_i \tau_i | \sigma_1 \tau_1^+ | n_k l_k j_k \tau_k > \; = \; (-1)^{l_k+j_i+3/2} \sqrt{6(2j_i+1)(2j_k+1)} \left\{ \begin{array}{ccc} 1/2 & l_k & j_k \\ j_i & 1 & 1/2 \end{array} \right\} \delta_{n_i n_k} \delta_{l_i l_k} \tag{11}$$

In the above, the curly bracket represents six j symbol. The above matrix element is nonzero only if the bra-side corresponds to a proton and the ket-side to a neutron i.e.$\tau_i = 1$ and $\tau_k = -1$

3. Results and Discussions

We apply the DSM formalism to the $2\nu\beta\beta$-decay of ^{76}Ge. We assume ^{56}Ni to be the inert core. The chosen configuration space is same as that used in extensive shell model calculation carried out for ^{76}Ge by Caurier et al[5], i.e. the model comprises $1p_{3/2}$, $0f_{5/2}$, $1p_{1/2}$ and $0g_{9/2}$ single particle orbitals. The spherical single particle energies of these orbitals relative to ^{56}Ni are taken as 0.0, 0.78, 1.08 and 4.5 MeV, respectively. The same spherical single particle energies are used both for protons as well as for neutrons. The effective interaction matrix elements used here have been given by Kuo and subsequently modified by Ahalpara et al.[4]. This effective interaction has been quite successfully used by us in describing many important features of the nuclei in this region.

The lowest prolate Hartree-Fock single particle spectrum for ^{76}Ge, ^{76}As and ^{76}Se are shown in Fig. 1, 2 and 3. In all the three nuclei, the protons are distributed entirely in the p-f single particle orbits. Four neutrons occupy g-orbit in each of the above nuclei. In ^{76}Ge, there is a well defined gap above the neutron Fermi surface. Hence single particle excitations over this gap would correspond to highly excited states of this nucleus and hence such excitations are not considered. However, very near the proton Fermi surface, there exists a $k = 3/2^-$ deformed orbital. By making 2p-2h excitations to this orbital from the valence $k = 1/2^-$ proton orbital and performing a tagged HF calculation, we get an excited intrinsic state which lies only 0.18 MeV above the lowest HF intrinsic state. There are many relatively high lying configurations which should ideally be considered while making a full scale spectroscopic calculation for this nucleus. However, the aim of our calculation is to perform model calculation for the same configuration space as used in shell model $2\nu\beta\beta$-decay study of ^{76}Ge [5] for which we have well established two-body interaction. As discussed in our earlier publication [6], the energy spectrum obtained by projecting out good angular momentum states from the lowest two intrinsic states and then performing band mixing calculation, are in good agreement with the experimental data for the lowlying levels.

An analysis of the HF single particle spectrum of ^{76}Se shows that there is well defined gap of about 3 MeV above the proton Fermi surface. Hence single-particle proton excitations would lie high in energy. However the gap above the neutron Fermi surface is relatively smaller. We considered an excited intrinsic state for this nucleus by

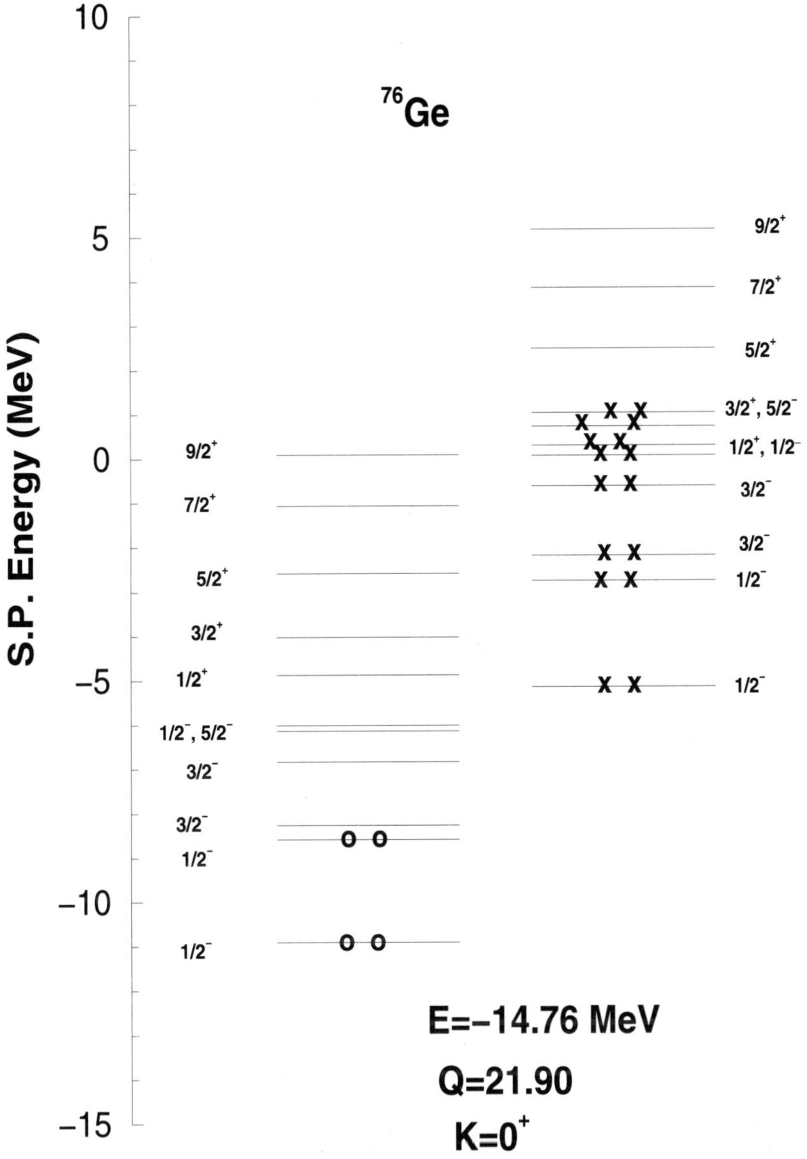

Figure 1. The lowest prolate HF single-particle spectra of ^{76}Ge

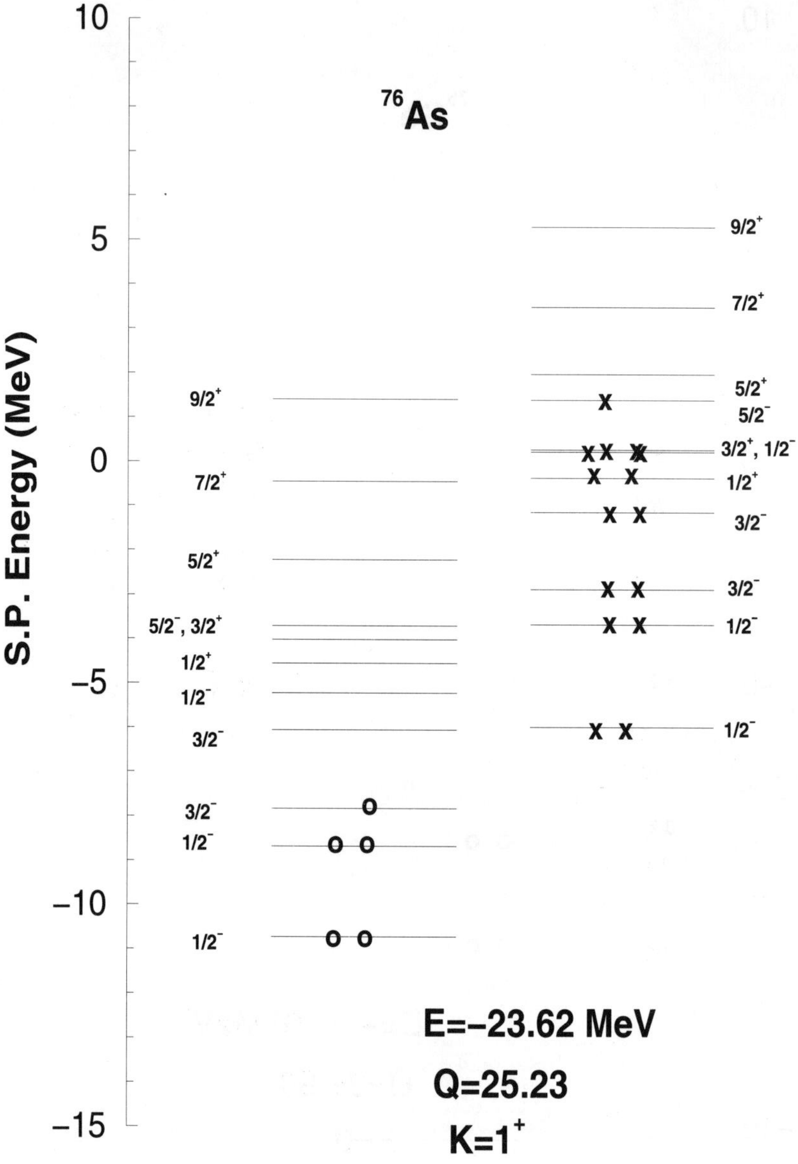

Figure 2. The lowest prolate HF single-particle spectra of ^{76}As

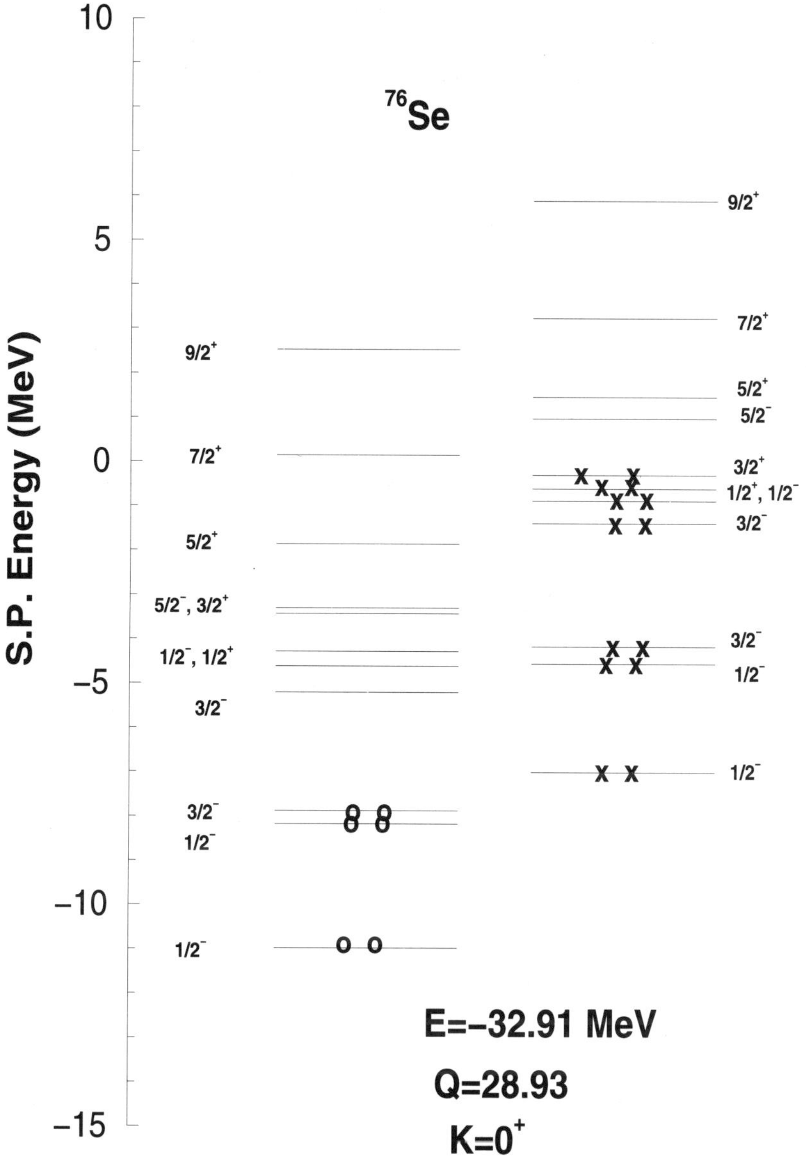

Figure 3. The lowest prolate HF single-particle spectra of ^{76}Se

promoting two neutrons from the valence $k = 1/2^-$ orbital to the unoccupied $k = 5/2^-$ orbital and then performing a tagged HF. This intrinsic state is about 1.4 MeV higher than the lowest one. Good angular momentum states are projected from each of these two intrinsic states and then a band mixing calculation is performed. Since the two intrinsic states are rather widely separated, the mixing in the band mixing calculation is rather small, about 15 %. The energy spectrum obtained from the band mixing calculation compares nicely with experiment[6].

The lowest $K = 1^+$ HF single particle spectrum for ^{76}As is shown in Fig. 2. There is no time reversal symmetry for this nucleus. Hence the deformed single particle energy of the state $|k>$ and its time reversal state $|-k>$ are not degenerate. However, in actual practice, the difference in the two energies are slight and we have ignored them in drawing the figure and showing the occupancies of the deformed states. However, in actual computation, the exact energies are taken into account. We consider all the $K = 1^+$ bands of this intermediate nucleus explicitly in the calculation of the Gamow-Teller matrix element for double beta decay. We have generated 17 $K = 1^+$ intrinsic states by particle-hole excitations. Good angular momentum states are projected from each of these intrinsic states. These good angular momentum states enter in the calculation of $2\nu\beta\beta$-decay matrix element in Eq. (1). Since the DSM formalism provides a good description of the low-lying spectroscopic properties of ^{76}Ge and ^{76}Se, the calculated wave functions gives us sufficient confidence. However DSM is unable to reproduce correctly the relative energies of the three nuclei. The main reason is the neglect of Coulomb energy. The DSM was mainly used for studying the spectroscopic properties of medium heavy nuclei and Coulomb energy was never important. One way of taking into account the Coulomb energy is to raise the spherical proton single particle energies relative to neutron orbits. We have found that if we raise the proton single particle energies by 8.1 MeV in each nucleus, then we get more realistic relative position of the above considered three states.

Assuming closure approximation and taking the experimental energies of the 0^+ and 1^+, the calculated results are given in Table 1. The agreement with experiment is quite satisfactory for the $0^+ \rightarrow 0^+$ transition. However for the transition $0^+ \rightarrow 2^+$ transition, only the lower limit of the half life has been experimentally measured. The calculated half life for this transition is larger than the value quoted by Dhiman and Raina [7] by about two orders of magnitude. More sophisticated experiments should be performed to measure the half life of this transition.

Table 1. Calculated $2\nu\beta\beta$ decay matrix elements and half lives for $0^+ \rightarrow 0^+$ and $0^+ \rightarrow 2^+$ transitions compared with experimental data.

Transitions	$M_{GT}^{2\nu}$	Calculated half life	Expt. half life
$0^+ \rightarrow 0^+$	0.144	1.41×10^{21}	1.3×10^{21}
$0^+ \rightarrow 2^+$	0.017	2.74×10^{25}	$>6.3 \times 10^{20}$

4. Conclusion

The deformed Shell model, which describes correctly the spectroscopy of of the low-lying excited states, has been applied for the calculation of the $2\nu\beta\beta$-decay of ^{76}Ge. The calculations have been performed for the model space used in extensive shell model calculations [5] with NN-interaction given by Kuo and modified by Ahalpara et al [4]. Our calculation reasonably reproduces the GT matrix elements and the half lifes.

Acknowledgement

This work is supported by Department of Science and Technology, Government of India.

References

[1] A. Faessler and F. Šimkovic, J. Phys. G **24**, 2139 (1998).

[2] S.R. Elliot and P. Vogel, Annu. Rev. Nucl. Part. **52**, 115 (2002).

[3] Y. Zdesenko, Rev. Mod. Phys. **74**, 663 (2002).

[4] D.P. Ahalpara, K.H. Bhatt and R. Sahu, J. Phys. G **11**, 735 (1985).

[5] E. Caurier, F. Nowacki, A. Poves and J. Retamosa, Phys. Rev. Lett. **77**, 1954 (1996).

[6] R. Sahu, F. Šimkovic and A. Faessler, J. Phys. G **25**, 1999 (1159).

[7] S.K. Dhiman and P.K. Raina, Phys. Rev. **50**, R2660 (1994)

NEUTRINOLESS DOUBLE BETA DECAY
Editors: V.K.B. Kota and U. Sarkar

POSITRON DOUBLE BETA DECAY ($2\beta^+/\beta^+EC/2EC$): CHALLENGES AND FUTURE PROSPECTS

P. K. Raina[1]

Indian Institute of Technology, Kharagpur 721302; India

Abstract

Positron double beta decay experiments have not been so popular in comparison to electron double beta decay. This article goes in to reasons for such a trend and also presents its importance in future course of activities in neutrinoless double beta decay out lining the challenges and future prospects.

1. Introduction

A brief introduction, importance and future prospects of neutrinoless double beta decay have already been discussed in the introductory article of this book. Further, the articles by Utpal Sarkar, V. K. B. Kota and P. K. Rath have already covered in details

[1] **e-mail: pkraina@phy.iitkgp.ernet.in**

the theoretical aspects of Neutrinoless double beta decay in exploring the physics beyond standard model (BSM) of fundamental particles. This article shall form a better reading if followed after these in a sequence. Here I concentrate on discussion of the positron double beta decay both for ($2\nu\beta\beta$) and ($0\nu\beta\beta$) decay modes.

1.1. Positron Double Beta decay

Double beta ($\beta\beta$) decay (DBD) mode being a second order weak interaction processes is one of the rare process of nature that takes place between two even-even nuclei. It can occur in two different modes. The two neutrino double beta (2ν $\beta\beta$) decay and the neutrinoless double beta (0ν $\beta\beta$) decay (NDBD). Both the modes can further be subdivided into two categories. The first one which takes place only as double electron $\beta^-\beta^-$ (e^-DBD) emission by converting two neutrons of a nucleus into two protons and two electrons having probability of emitting or not emitting the two neutrinos. The DBD with emission of two neutrinos has been detected in many experiments for nearly a dozen of nuclei in past two decades for e^-DBD with half lives ranging between 10^{18} to 10^{21} yrs. First successful direct measurement was made in 1987 after half a century of trials which were dominated by geochemical and radiochemical processes in the begining. This is a big achievement of last two decades that has opened up very challenging experimentation field of NDBD [1, 2, 3, 4]. As discussed in the introductory article, the tremendous potential of this area of research has been realized globally.

The second category of DBD involves nuclei where two protons of a nucleus get converted to two neutrons with emission of two positrons. Since positron has good probability for capture of atomic electron, two additional options open up depending on capture of one electron (β^+EC) or two electrons ($2EC$). Thus for the same nucleus we have three competing processes ($2\beta^+$, β^+EC and $2EC$ and shall refer to these as e^+DBD. The DBD modes can be represented as

1. *($2\nu\beta\beta$) decay mode*

(a) Double Electron emission (A, Z) $\rightarrow$ (A, Z+2) + $2e^-$ + $2\overline{\nu}_e$.
(b) Double Positron emission (A, Z) $\rightarrow$ (A, Z-2) + $2e^+$ + $2\nu_e$.
(c) Positron emission/Electron capture e^- + (A, Z) $\rightarrow$(A, Z-2) + e^+ + $2\nu_e$.
(d) Double electron capture $2e^-$ + (A, Z) $\rightarrow$ (A, Z-2) + $2\nu_e$.

2. *($0\nu\beta\beta$) decay mode*

(a) Double Electron emission (A, Z) $\rightarrow$ (A, Z+2) + $2e^-$
(b) Double Positron emission (A, Z) $\rightarrow$ (A, Z-2) + $2e^+$
(c) Positron emission/Electron capture e^- + (A, Z)$\rightarrow$ (A, Z-2) + e^+
(d) Double electron capture $2e^-$ + (A, Z) $\rightarrow$ $(A, Z-2)^*$

1.2. Experimental Challenges and Their Implications

All the possible e^+DBD candidates allowed energetically are listed in table 1 [5]. The possible competing modes, Q values for the decay and abundances for parent nuclei are also included so as to visualize the situation for possible experimental candidates. The experimental sensitivity in searching for double beta decay gets limited for $\beta^-\beta^-$ decay mode because of the presence of electron background. Thus one would like to explore some alternative way for investigating the interesting physics of double beta decay. The Study of other possible modes ($2\beta^+/\beta^+EC/2EC$) are relatively easier from the standpoint of separability from the background contaminations. Although the low Q-values for ($2\beta^+/\beta^+EC$) transitions have been a factor which have hindered experiments but these transitions are attractive from the experimental view due to the possibility to detect either the coincidence signals from four/two annihilation γ-rays from two/one positrons.

On the other hand, the Q-value for the $2EC$ process can be large enough (up to $\approx$ 2.8 MeV) but the detection of the ground state transition is made difficult since only X-rays are emitted. But there is possibility of $2EC$ process with the excited final state which allows a better detection due to emission of correlated de-excitation photon. On the other hand, β^+EC process half life gets reduced in comparison to $2\beta^+$ decay for larger Q value and lesser coulomb repulsion but has an advantage over $2EC$ due to the emission of a pair of coincident photons. Experimental investigations of these modes have revived recently [6, 7, 8, 9, 10]. Once the $0\nu\beta\beta$ decay is observed even for e^-DBD, the e^+DBD processes would play a crucial role in discriminating the finer issues [11, 12, 13] like dominance of Majorana neutrino mass or the right handed current contributions.

Experimental searches for e^+DBD processes are not so popular as those for e^-DBD. The two major distinctions as can be seen from table 1 are that

(i) the e^-DBD has only one mode $2\beta^-$ where as e^+DBD has three modes: $2\beta^+$, mixed β^+EC and $2EC$ mode.

(ii) the large difference in natural abundance (δ) which is very crucial consideration from both large isotope mass and maximum radiopurity requirement. One observes from tables of DBD that 30 out of 35 nuclei for e^-DBD have natural abundance more than 5% where as only 6 nuclei from full list of 34 e^+DBD cases have δ greater than 5%.

In case of $2\beta^+$ decay, energy release should exceed 4 electron masses, and only 6 nuclei from the full list of 34 cases have enough energy for this. All these nuclei can undergo also β^+/EC decays (for which energy release should exceed $2m_ec^2 + E_b$, where E_b is binding energy of captured electron on atomic shell), together with 22 additional nuclides. And all of them, together with 12 other nuclei (^{36}Ar, ^{40}Ca, ^{54}Fe, ^{108}Cd, ^{126}Xe, ^{132}Ba, ^{138}Ce, ^{152}Gd, ^{158}Dy, ^{164}Er, ^{180}W, and ^{196}Hg) can undergo $2EC$ capture process.

Energy releases in $2\beta^+$ and β^+EC decays are lower than those in e^-DBD that result in lower probabilities for such processes due to suppression by small phase space factors. For neutrinoless processes, because of relation between half-life, phase space factor G, nuclear matrix element (NME) and an effective value of the electron neutrino mass $\langle m_\nu \rangle$: $[T_{1/2}]^{-1} = G \cdot |\mathrm{NME}|^2 \cdot \langle m_\nu \rangle^2$, lower G value for $2\beta^+$ and β^+/EC processes will result in worse limits on neutrino mass (or other theoretical parameters), even if the same experimental $T_{1/2}$ limit will be reached as that for e^-DBD. For these reasons there has been lower popularity of searches for e^+DBD processes.

Table 1. List of all possible naturally occuring positron double beta ($2\beta^+/\beta^+EC/2EC$) decaying nuclei. The Q values of ($0^+ \to 0^+$) decay and the natural abundances of the parent isotopes are taken from [5].

Nuclear Transition	Possible decay Channels	$Q_{\beta\beta}$ (keV)	Natural abundance(%)
$^{36}Ar_{18} \to ^{36}S_{16}$	$2EC$	433.5±0.4n	0.3365±0.003
$^{40}Ca_{20} \to ^{40}Ar_{18}$	$2EC$	193.78±0.29	96.94±0.156
$^{50}Cr_{24} \to ^{50}Ti_{22}$	$\beta^+EC/2EC$	1171.4±1.2	4.344±0.013
$^{54}Fe_{26} \to ^{54}Cr_{24}$	$2EC$	679.9±0.5	5.844±0.035
$^{58}Ni_{28} \to ^{58}Fe_{26}$	$\beta^+EC/2EC$	1925.8±0.7	68.0769±0.0089
$^{64}Zn_{30} \to ^{64}Ni_{28}$	$\beta^+EC/2EC$	1096.4±0.9	48.63±0.60
$^{74}Se_{34} \to ^{74}Ge_{32}$	$\beta^+EC/2EC$	1209.4±0.6	0.89±0.04
$^{78}Kr_{36} \to ^{78}Se_{34}$	$2\beta^+/\beta^+EC/2EC$	2866.0±7.0	0.35±0.01
$^{84}Sr_{38} \to ^{84}Kr_{36}$	$\beta^+EC/2EC$	1786.8±3.6	0.56±0.01
$^{92}Mo_{42} \to ^{92}Zr_{40}$	$\beta^+EC/2EC$	1649.1±3.7	14.84±0.35
$^{96}Ru_{44} \to ^{96}Mo_{42}$	$2\beta^+/\beta^+EC/2EC$	2719±8	5.54±0.14
$^{102}Pd_{46} \to ^{102}Ru_{44}$	$\beta^+EC/2EC$	1172.0±2.6	1.02±0.01
$^{106}Cd_{48} \to ^{106}Pd_{46}$	$2\beta^+/\beta^+EC/2EC$	2771±8	1.25±0.06
$^{108}Cd_{48} \to ^{108}Pd_{46}$	$2EC$	269±6	0.89±0.03
$^{112}Sn_{50} \to ^{112}Cd_{48}$	$\beta^+EC/2EC$	1922±4	0.97±0.01
$^{120}Te_{52} \to ^{120}Sn_{50}$	$\beta^+EC/2EC$	1698±10	0.09±0.01
$^{124}Xe_{54} \to ^{124}Te_{52}$	$2\beta^+/\beta^+EC/2EC$	2865.6±2.2	0.09±0.01
$^{126}Xe_{54} \to ^{126}Te_{52}$	$2EC$	897±6	0.09±0.01
$^{130}Ba_{56} \to ^{130}Xe_{54}$	$2\beta^+/\beta^+EC/2EC$	2611±7	0.106±0.001
$^{132}Ba_{56} \to ^{132}Xe_{54}$	$2EC$	839.9±3.2	0.101±0.001
$^{136}Ce_{58} \to ^{136}Ba_{56}$	$2\beta^+/\beta^+EC/2EC$	2400±50	0.185±0.002
$^{138}Ce_{58} \to ^{138}Ba_{56}$	$2EC$	693±10	0.251±0.002
$^{144}Sm_{62} \to ^{144}Nd_{60}$	$\beta^+EC/2EC$	1781.1±1.8	3.07±0.07
$^{152}Gd_{64} \to ^{152}Sm_{62}$	$2EC$	55.6±1.2	0.20±0.01
$^{156}Dy_{66} \to ^{156}Gd_{64}$	$\beta^+EC/2EC$	2011±6	0.06±0.01
$^{158}Dy_{66} \to ^{158}Gd_{64}$	$2EC$	283.3±2.4	0.10±0.01
$^{162}Er_{68} \to ^{162}Dy_{66}$	$\beta^+EC/2EC$	1844.5±2.8	0.14±0.01
$^{164}Er_{68} \to ^{164}Dy_{66}$	$2EC$	24.1±2.5	1.61±0.03
$^{168}Yb_{70} \to ^{168}Er_{68}$	$\beta^+EC/2EC$	1422.1±3.7	0.13±0.01
$^{174}Hf_{72} \to ^{174}Yb_{70}$	$\beta^+EC/2EC$	1101.1±2.3	0.16±0.01
$^{180}W_{74} \to ^{180}Hf_{72}$	$2EC$	146±5	0.12±0.01
$^{184}Os_{76} \to ^{184}W_{74}$	$\beta^+EC/2EC$	1441.5±1.4	0.02±0.01
$^{190}Pt_{78} \to ^{190}Os_{76}$	$\beta^+EC/2EC$	1383±6	0.014±0.001
$^{196}Hg_{80} \to ^{196}Pt_{78}$	$2EC$	819.7±3.0	6.87±0.15

2. Experimental Investigations of e^+DBD Processes

From total of 34 e^+DBD nuclei, the direct experimental limits on $T_{1/2}$ have been put only for 21 nuclei. The best claimed limit so far is at the level of $\simeq 10^{21}$ yr which could be compared with limits on the level of $10^{23} - 10^{25}$ yr established for several e^-DBD nuclides. Recently observation of e^+DBD are reported for ^{64}Zn and ^{130}Ba. However, there are some ambiguities between different results. Although there have been attempts for many cases looking at option of three competing processes of e^+DBD, details about current status for three more promising nuclei for future are discussed below. Exhaustive data of available results can be found in data tables [14, 15] and an updated review on e^+DBD is in the process of writing that shall be available in near future.

2.1. *Current Status*

There is a claim on observation of the β^+EC decay for ^{64}Zn. The emitted positrons in this case have quite low energy of ≤ 74 keV and makes it difficult to extract any useful limits [16]. Other option of finding the characteristic X rays after $2EC$ capture were searched [17] and the limit is set at the level of 10^{16} yr. In 1985, β^+EC decay of ^{64}Zn was attempted with two HP Ge detectors of 110 cm^3 looking for 511 keV gamma rays in coincidence that could achieve the limit of $T_{1/2}^{0\nu+2\nu} > 2.3\times^{18}$ yr [18]. Further doubts were raised since the experiment was performed at the earth level and possible sources which can imitate the effect were not analyzed satisfactorily. Preliminary results of another recent (2003) measurements (HP Ge 456 cm^3 and CsI(Tl) $\simeq$400 cm^3 in coincidence, 460 g of Zn sample, 375 h of measurements underground in the Cheong Pyung Laboratory, depth of 1000 m w.e.) [6] refute observation with such a rate giving only a limit $T_{1/2}^{0\nu+2\nu} > 1.3\times^{20}$ yr. However, the results are not published as yet. The best current limits on $2EC$ decays of ^{64}Zn are from two experiments: (i) 2003, semiconductor detector $Cd_{0.9}Zn_{0.1}Te$, mass of 2.89 g, 1117 h of data collection, half-life limits on the level of 10^{17} yr [19] and (ii) 2005, $ZnWO_4$ crystal scintillator, mass of 4.5 g, 429 h of data collection in the Solotvina Underground Laboratory (1000 m w.e.), half-life limits on the level of 10^{18} yr [20]. Very recently [21] data of $ZnWO_4$ crystall scintillator of mass 117g kept in LNGS for total measurement time of 1902 h with 90% C.L. has put the half life limits at $4.0\times^{18}$ y and $2.1\times^{20}$ y for 2EC 2K and β^+/EC for $2\nu\ \beta\beta$ decay. In case of $0\nu\ \beta\beta$ decay the limits have been put as $4.0\times^{18}$ y for 2EC 2K, $3.4\times^{18}$ y for 2EC and $2.2\times^{20}$for β^+/EC modes.

For ^{106}Cd nucleus the $2\beta^+$ decay is also energetically allowed and is one of the most investigated e^+DBD nuclides. The data collected during 2982 h in an experiment with a small $^{116}CdWO_4$ crystal scintillator (15 cm^3, enriched in ^{116}Cd to 83%), performed in the Solotvina Underground Laboratory (1000 m w.e.), were also used to set limits for ^{106}Cd on the level of 10^{17} to 10^{19} yr [23], independent of mode of decay. A "source = detector" approach and low energy threshold allowed for the first time to set a limit on $2EC$ at ($\simeq 10^{17}$ yr). Results of two experiments with ^{106}Cd were published in 1996. In the first one [24], the external Cd foil (331 g) was measured in the Frejus Underground Laboratory (4800 m w.e.) with the 120 cm^3 HP Ge detector during 1137 h; gamma quanta from annihilations of positrons and from de-excitations of the daughter ^{106}Pd nucleus (in case if the ^{106}Pd excited levels were populated) were searched for. Half-life limits on the level of 10^{18}–10^{19} yr were reached. In the second one [25], a large (1.046 kg) $CdWO_4$ scintillator with natural Cd composition was measured in the Gran Sasso Underground Laboratory (3600 m w.e.) during 6701 h. Determined limits on half-life for $2\beta^+$ and β^+EC decays were on the level of $\simeq 10^{19}$ yr for 0ν, and $\simeq 10^{17}$ yr for 2ν processes. In 2003, the long-term (14183 h) experiment in the Solotvina Underground Laboratory with 3 enriched $^{116}CdWO_4$ scintillators (total mass of 330 g) and 433 h data with the non-enriched $CdWO_4$ crystal of 454 g improved [26] sensitivity of the experiment near one order of magnitude. To overcome this drawback, in 1999 experiment [31] with an enriched ^{106}Cd to 68% was used (total mass of 154 g). Measurements in the Gran Sasso Underground Laboratory with two low background NaI(Tl) scintillators during 4321 h gave possibility to reach level of sensitivity of more than 10^{20} yr for $2\beta^+$, β^+EC and 2EC processes which resulted in emission of γ quanta. These are currently the world best values [22].

There are two running experiments currently to probe the 2β decays of ^{106}Cd: COBRA and TGV-II. In the COBRA experiment [19] (Gran Sasso National Laboratory), two small detectors were measured: $Cd_{0.9}Zn_{0.1}Te$ (mass of $\simeq$3 g) during 1117 h, and CdTe ($\simeq$6 g) during 1645 h. While these detectors are semiconductors, their energy resolution is far from typical for Ge detector: at energy of 1000 keV FWHM is equal $\simeq$32 keV for CdZnTe, and $\simeq$57 keV for CdTe crystals. Results for ^{106}Cd $T_{1/2}$ limits were in the range of 10^{17}-10^{18} yr. Sensitivity of the COBRA experiment should be improved in future with bigger number of detectors and with crystals of bigger mass. In the TGV-II experiment [27] (Modane Underground Laboratory), 32 planar HP Ge detectors are used. The Cd foils, inserted between neighbouring detectors, are enriched in ^{106}Cd to 75%. After 3736 h of data taking, limits on 2EC decays of ^{106}Cd to the ground state and excited levels of ^{106}Pd are higher than 10^{19} yr.

^{130}Ba is also one of the 6 potentially $2\beta^+$ decaying nuclei. In 1996, the data of the old geochemical measurements on amount of daughter nuclide ^{130}Xe accumulated in $BaSO_4$ sample during geological times, the limit of $T_{1/2} > 4 \times 10^{21}$ yr [28] for all modes of ^{130}Ba decays was determined(geochemical measurements can give only amount of daughter nuclide but are unable to distinguish the mode of decay: $2\beta^+$ or β^+EC or 2EC as well as 0ν or 2ν). In 2001, a claim on positive observation of the ^{130}Ba decay followed from geochemical measurements of ^{130}Xe excess in $BaSO_4$ sample; the derived half-life was $T_{1/2} = (2.16 \pm 0.52) \times^{21}$ yr [29], for all modes of decay. This is in a contradiction with the above-mentioned limit of $T_{1/2} > 4 \times 10^{21}$ yr. If confirmed, it would be the first observation of the e^+DBD. $T_{1/2}$ value is consistent with theoretical expectations for 2EC capture as seen in theoretical calculations. Recently efforts have succeeded [7] in measurements of BaF_2 crystal scintillator of mass 3615 g and with two low background NaI(Tl) detector at LNGS and half limit at 90% C.L. have been achived for g.s to g.s. and excited state in the order of 10^{17} y.

2.2. Some Possible Experiments in Near Future

In this section the current planed approaches for above three nuclei are discussed for near future that can improve on measurement limits in the e^+DBD from technicality of an active detector and another important issue of radiopurity and economics angle.

Search for double beta processes in ^{64}Zn

This nucleus has maximum natural isoptopic abundance. An experiment to search for e^+DBD processes in ^{64}Zn (energy release = 1095.7 keV, natural isotopic abundance = 48.268%) at the level of sensitivity $\sim (1-5) \times 10^{21}$ yr with the help of $ZnWO_4$ (0.12 Kg) crystal scintillators is completed recently [21] in the low-background DAMA/R & D set-up in the Gran Sasso National Laboratory of INFN (Italy). This sensitivity is close the controversial result for $\beta^+ EC/2EC$ decay with $T_{1/2} = (1.1 \pm 0.9) \times 10^{19}$ yr [30]. Primarily this is an R & D proposal to check energy resolution, background conditions, ability to discriminate PMT noise to explore low energy region where 2EC process is expected. This has helped to put a larger detector with mass up to 0.7 kg that is in progress and is good enough to check the controversial result.

Proposal of ^{106}Cd experiment

The ^{106}Cd isotope is a one of the best candidate nuclides because of its rather high Q value (2770 ± 7) keV and to the favourable theoretical estimates for half-lives. To reach a suitable sensitivity, the use of enriched samples is needed to effectively exploit the potentiality of the 2EC, $\beta^+ EC$ and $2\beta^+$ processes. Approximately 115 g of enriched ^{106}Cd isotope available in the Institute for Nuclear Research (Kiev, Ukraine) can be used in an experiment to search for 2EC, $\beta^+ EC$ and $2\beta^+$ processes in ^{106}Cd at the level of sensitivity $10^{21} - 10^{22}$ yr. Taking into account theoretical predictions double beta decay of ^{106}Cd could be detected at this level of sensitivity. Moreover, this feature allows to realise experiment also by the passive source approach, searching for the emitted γ-rays. This experiment can follow the same approach as the one used by the DAMA collaboration [31]. The main radioactive residual pollution in the set-up was due to radioactive contamination of the cadmium samples: (0.13 ± 0.01) mBq/g for ^{232}Th, (0.12 ± 0.04) mBq/g for ^{226}Ra and < 2 mBq/g for ^{40}K. Therefore, a higher sensitivity can be obtained by using enriched cadmium samples with a better purification. Higher efficiency in an experiment can be reached by using radiopure detectors with very high registration efficiency to γ quanta (for instance $CdWO_4$ crystal scintillators) without any assembling materials between the samples and detector. Very competitive results can be reached with the growth of $^{106}CdWO_4$ scintillator(s), which could be used as source-detector(s). Similar experiments have been already realised using $^{116}CdWO_4$ scintillators to study the $2\beta^-$ decay of ^{116}Cd. An experiment with ^{116}Cd has been performed by the INR (Kiev) at the Solotvina Underground Laboratory (Ukraine) with four $^{116}CdWO_4$ crystal scintillators (total mass 339 g) enriched to 83% in ^{116}Cd.

In the search for β^+EC/EC array of $^{106}CdWO_4$ detectors can be considered to search for multiple coincidences due to the 511 keV γ-rays generated by the annihilation of the emitted positron(s), with considerable reduction of the measured background. Sensitivity of the experiment to 2EC process (the decay channel with the lowest theoretically estimated half-life) can be significantly improved by depletion of ^{113}Cd in the enriched ^{106}Cd by 4-5 order of magnitude. Such a depletion procedure can be realized without noticeable losses of high cost enriched ^{106}Cd.

Search for double beta processes using rare gases detectors

Two of the six $2\beta^+$ potentially decaying nuclides are noble gas isotopes ^{124}Xe and ^{78}Kr. They have the highest available energies which could be released in positron double beta decay processes (2864 keV for ^{124}Xe and 2846 keV for ^{78}Kr) and can be employed to realize experiments with the "active source" method. Unfortunately these nuclides have low natural abundances ($\delta = 0.0952\%$ for ^{124}Xe and $\delta = 0.355\%$ for ^{78}Kr) and require an expensive procedure for the isotopical enrichment.

Some possible future experimental configurations(Xe) will employ ~ 100 kg of LXe, enriched in ^{124}Xe, in coincidence with high Z low background solid scintillators [32] . The reachable sensitivity is of the order of $10^{25} - 10^{26}$ yr for $0\nu e^{+}DBD$ processes.

As regards ^{78}Kr, in 1994 a preliminary experiment using 35 l of enriched (to 94.08%) ^{78}Kr obtained $T_{1/2}$ limits for $\beta^{+}EC$ and $2\beta^{+}$ processes, which are among the best reached for 2EC, $\beta^{+}EC$ and $2\beta^{+}$ nuclei ($\sim 10^{21}$ yr). Data were collected in the Canfranc Tunnel Laboratory (675 m.w.e.) during 4435 h, with a high pressure ionization chamber placed within an array of 6 large NaI scintillators. NaI scintillators were used to measure in coincidence the 2 (or 4) γ quanta of 511 keV after annihilation of 1 (or 2) emitted positrons, while ionization chamber measured the energy of emitted β^{+} particle(s).

As regards proposals for future experiments, here we just remind ref. [33], where an "emission" detector, self-triggered with scintillation, has been proposed to search for positron double beta decays of ^{124}Xe or ^{78}Kr. Total efficiency of the detector is calculated to be $\sim 60\%$. Proposed detector with 0.4 m^3 of liquid Krypton or Krypton-Xenon(3%) mixture with Krypton enriched in ^{78}Kr can be able to reach a sensitivity of $\sim 10^{27}$ yr and to check the theoretical prediction for $T_{1/2}$ of $\beta^{+}EC$ and $2\beta^{+}$ decay mode of ^{78}Kr. Main difficulties in the realisation of this experiment are connected to the expensive cost to obtain ~ 1 t of enriched ^{78}Kr and its purification, to the realization of a cryogenic system and to the monitoring of the LKr working condition during long period of data taking.

Further motivation for e^{+}DBD searches has been furnished because of the only way to get information about the 0ν double beta decays particle physics parameters $\langle m_\nu \rangle$, $\langle \lambda \rangle$ and $\langle \eta \rangle$ is from a simultaneous analysis of a high sensitive $2\beta^{-}$ experiment (which has observed $2\beta^{-}0\nu$ decay as ^{76}Ge) and a suitable very high-sensitive mixed-mode experiment looking for 0ν $\beta^{+}EC$ decay.

3. Nuclear Structure Issues

The difficulties in extracting information on the neutrino properties of these processes relates to the strong dependence of the event rates on nuclear properties via nuclear matrix elements of various weak-interaction processes involved. The $2\nu\beta\beta$ decay, allowed in SM, helps to some extent in checking the validity of different nuclear models and provides important information about the nuclear structure. The Half-life of $2\nu\beta\beta$ decay is a product of an accurately known phase factor and an appropriate nuclear transition matrix element $M_{2\nu}$ being calculated in various nuclear models. The calculation of $M_{2\nu}$ involves sum over virtual intermediate nuclear states and is a challenging many-body problem. The values of nuclear transition matrix element $M_{2\nu}$ can be extracted directly from the nucleus for which half-life has been measured. Consequently, the validity of different models employed for nuclear structure calculations can be tested by calculating $M_{2\nu}$. The effective majorana neutrino mass depends on the

Table 2. Experimental limits on half-lives $T_{1/2}^{2\nu}(0^+ \to 0^+)$, theoretically calculated $M_{2\nu}$ and corresponding $T_{1/2}^{2\nu}(0^+ \to 0^+)$ for 2ν $\beta^+\beta^+$, 2ν β^+EC and 2ν ECEC decay of ^{96}Ru, ^{102}Pd, ^{106}Cd and ^{108}Cd nuclei. Half-lives are calculated using g_A= (1.261 - 1.0).

Nuclei	Decay	Experiment		Theory			
	Mode	Ref	$T_{1/2}^{2\nu}$ (y)	Ref.	Model	$\lvert M_{2\nu}\rvert$	$T_{1/2}^{2\nu}$ (y)
^{96}Ru	$\beta^+\beta^+$	[18]	$> 3.1\times10^{16*}$	Present	PHFB	0.0537	$(1.378\text{-}3.485)\times10^{28}$
				[12]	QRPA	0.2510	$(6.309\text{-}15.950)\times10^{26}$
	β^+EC	[18]	$> 6.7\times10^{16*}$	Present	PHFB	0.0537	$(3.599\text{-}9.100)\times10^{23}$
				[34]	SU(4)$_{\sigma\tau}$	0.1005	$(1.028\text{-}2.598)\times10^{23}$
				[12]	QRPA	0.2694	$(1.430\text{-}3.616)\times10^{22}$
	ECEC		-	Present	PHFB	0.0537	$(0.644\text{-}1.628)\times10^{23}$
				[34]	SU(4)$_{\sigma\tau}$	0.1005	$(1.839\text{-}4.649)\times10^{22}$
				[12]	QRPA	0.2620	$(2.705\text{-}6.840)\times10^{21}$
^{102}Pd	β^+EC			Present	PHFB	0.0524	$(2.509\text{-}6.344)\times10^{32}$
	ECEC			Present	PHFB	0.0524	$(3.783\text{-}9.565)\times10^{24}$
^{106}Cd	$\beta^+\beta^+$	[35]	$>5.0\times10^{18}$	Present	PHFB	0.0819	$(3.495\text{-}8.836)\times10^{27}$
		[31]	$>2.4\times10^{20*}$	[36]	SQRPA(l.b.)	0.61	$(6.304\text{-}15.940)\times10^{25}$
		[37]	$>1.0\times10^{19*}$		SQRPA(s.b.)	0.57	$(7.220\text{-}18.260)\times10^{25}$
		[25]	$>9.2\times10^{17}$	[38]	QRPA(AWS)	0.722	$(4.500\text{-}11.380)\times10^{25}$
		[39]	$>2.6\times10^{17*}$		QRPA(WS)	0.166	$(8.513\text{-}21.520)\times10^{26}$
		[40]	$>6\times10^{16}$	[37]	QRPA(WS)	0.840	$(3.324\text{-}8.406)\times10^{25}$
					QRPA(AWS)	0.780	$(3.856\text{-}9.749)\times10^{25}$
				[12]	QRPA	0.218	$(4.936\text{-}12.480)\times10^{26}$
				[41]	QRPA		4.940×10^{25}
	β^+EC	[35]	$>1.2\times10^{18}$	Present	PHFB	0.0819	$(9.489\text{-}23.992)\times10^{22}$
		[31]	$>4.1\times10^{20}$	[36]	SQRPA(l.b.)	0.61	$(1.712\text{-}4.328)\times10^{21}$
		[37]	$>0.66\times10^{19*}$		SQRPA(s.b.)	0.57	$(1.960\text{-}4.957)\times10^{21}$
		[25]	$>2.6\times10^{17}$	[38]	QRPA(AWS)	0.718	$(1.236\text{-}3.124)\times10^{21}$
		[39]	$>5.7\times10^{17*}$		QRPA(WS)	0.168	$(2.257\text{-}5.706)\times10^{22}$
				[34]	SU(4)$_{\sigma\tau}$	0.1947	$(1.680\text{-}4.248)\times10^{22}$
				[42]	RQRPA(AWS)	0.56	$(2.031\text{-}5.136)\times10^{21}$
					RQRPA(WS)	0.55	$(2.106\text{-}5.324)\times10^{21}$
				[37]	QRPA(WS)	0.84	$(9.027\text{-}22.820)\times10^{20}$
					QRPA(AWS)	0.78	$(1.047\text{-}2.647)\times10^{21}$
				[12]	QRPA	0.352	$(5.141\text{-}13.000)\times10^{21}$
				[43]	QRPA(WS)	0.493	$(2.621\text{-}6.626)\times10^{21}$
						0.660	$(1.462\text{-}3.697)\times10^{21}$

Nuclei	Decay Mode	Experiment Ref	Experiment $T_{1/2}^{2\nu}$ (y)	Theory Ref.	Model	$\lvert M_{2\nu}\rvert$	$T_{1/2}^{2\nu}$ (y)
^{106}Cd	ECEC	[31]	$>7.3\times10^{19}$	Present	PHFB	0.0819	$(1.293\text{-}3.270)\times10^{22}$
		[8]	$>1.0\times10^{18}$	[36]	SQRPA(l.b.)	0.61	$(2.333\text{-}5.899)\times10^{20}$
		[19]	$>1.0\times10^{18}$		SQRPA(s.b.)	0.57	$(2.672\text{-}6.756)\times10^{20}$
		[23]	$>5.8\times10^{17}$	[38]	QRPA(AWS)	0.718	$(1.684\text{-}4.258)\times10^{20}$
		[35]	$>5.8\times10^{17}$		QRPA(WS)	0.168	$(3.076\text{-}7.780)\times10^{21}$
				[44]	SSDH(Theo)	0.28	$(1.107\text{-}2.800)\times10^{21}$
					SSDH(Exp)	0.17	$(3.004\text{-}7.595)\times10^{21}$
				[34]	SU(4)$_{\sigma\tau}$	0.1947	$(2.290\text{-}5.790)\times10^{21}$
				[42]	RQRPA(AWS)	0.56	$(2.768\text{-}6.999)\times10^{20}$
					RQRPA(WS)	0.55	$(2.870\text{-}7.256)\times10^{20}$
				[37]	QRPA(WS)	0.84	$(1.230\text{-}3.111)\times10^{20}$
					QRPA(AWS)	0.78	$(1.427\text{-}3.608)\times10^{20}$
				[12]	QRPA	0.270	$(1.191\text{-}3.011)\times10^{21}$
				[43]	QRPA(WS)	0.493	$(3.572\text{-}9.031)\times10^{20}$
						0.660	$(1.993\text{-}5.039)\times10^{20}$
^{108}Cd	ECEC	[35]	$>4.1\times10^{17}$	Present	PHFB	0.0952	$(3.939\text{-}9.959)\times10^{27}$
		[19]	$>1.0\times10^{18}$				
		[23]	$>4.1\times10^{17}$				
^{124}Xe	$\beta^+\beta^+$			Present	PHFB	0.0525	$(3.015-7.624)\times10^{27}$
		[45]	$>2.0\times10^{14}$	[12]	QRPA	0.077	$(1.4-3.54)\times10^{27}$
	β^+EC	[45]	$>4.8\times10^{16}$	Present	PHFB	0.0525	$(0.835-2.111)\times10^{23}$
				[34]	$SU(4)_{\sigma\tau}$	0.0528	$(0.82-2.073)\times10^{23}$
				[46]	MCM	0.01	$(2.30-5.81)\times10^{23}$
				[12]	QRPA	0.0875	$(3.0-7.59)\times10^{22}$
	ECEC	[47]	$>1.1\times10^{17}$	Present	PHFB	0.0525	$(0.712-1.801)\times10^{22}$
				[34]	$SU(4)_{\sigma\tau}$	0.0528	$(0.7-1.77)\times10^{22}$
				[46]	MCM	0.0071	$(3.9-9.8)\times10^{23}$
				[12]	QRPA	0.082	$(0.29-0.73)\times10^{22}$
^{126}Xe	ECEC			Present	PHFB	0.0487	$(0.568-1.437)\times10^{25}$
^{130}Ba	$\beta^+\beta^+$	[37]	$>4.0\times10^{21}$	Present	PHFB	0.0415	$(0.568-1.437)\times10^{28}$
		[29]	$2.2\times10^{21}{}_*$	[12]	QRPA	0.069	$(1.7-4.3)\times10^{29}$
	β^+EC			Present	PHFB	0.0415	$(0.419-1.059)\times10^{24}$
				[34]	$SU(4)_{\sigma\tau}$	0.056	$(2.2-5.56)\times10^{23}$
				[46]	MCM	0.015	$(3.1-7.83)\times10^{24}$
				[12]	QRPA	0.084	$(1.0-2.53)\times10^{23}$
	ECEC			Present	PHFB	0.0415	$(1.405-3.553)\times10^{22}$
				[34]	$SU(4)_{\sigma\tau}$	0.056	$(0.75-1.9)\times10^{22}$
				[46]	MCM	0.007	$(0.47-1.18)\times10^{24}$
				[12]	QRPA	0.075	$(0.42-1.06)\times10^{22}$
^{132}Ba	ECEC	[37]	$>3.0\times10^{20}$	Present	PHFB	0.0522	$(0.547-1.384)\times10^{25}$

accuracy with which matrix element can be calculated, hence nuclear structure effects in double beta decay are very important. This has stimulated important activities in nuclear structure theory, including Monte Carlo and Lanczos shell model techniques [48] and quasiparticle RPA [49]. Table 2 summarizes most of the results available for nuclei in mass region 96 to 132. Here theoretical results for ^{106}Cd and ^{130}Ba are more relevant from the point of discussion of presented in experimental section.

Several nuclear models have been employed to calculate the $2\nu\beta\beta$ and $0\nu\beta\beta$ decay transition matrix elements in two-nucleon (2n) mechanism. It has been observed that in all cases the $2\nu\beta\beta$ decay matrix elements are sufficiently quenched. Hence the main motive of all the theoretical calculations is to achieve the quenching of $M_{GT}^{2\nu}$ and to understand the physical mechanism responsible for quenching. Since the observation of $0\nu\beta\beta$ decay is controversial and has not been confirmed so far, the models predict half-lives assuming a certain value for the neutrino mass or conversely extract various parameters from the observed limits of half-lives of $0\nu\beta\beta$ decay.

The reliability of the predictions can be judged a priori only from the success of the model in explaining various observed physical properties of nuclei. Here I have presented a brief introduction along with some references to different nuclear models in use for DBD matrix elements including the recently studied ones by us within the Projected Hartree-Fock-Bogoliubov (PHFB) formalism.

Shell model and its extensions

Shell model attempts to solve the nuclear many-body problem as exactly as possible. It is quite successful for nuclei A $<$ 40. But for A $>$ 40, the number of basis states increase so drastically that except for ^{48}Ca it is impossible to perform a conventional shell model calculation without imposing certain truncation scheme. On the other hand most of the double beta decay emitters are medium or heavy mass nuclei. Hence a reliable shell model calculation is difficult to perform. Recent shell model calculations [50], [48] are more realistic in nature and better large-scale shell model calculations are becoming feasible with availability of computer facilities. The Shell Model Monte Carlo (SMMC) Method is a recent development over the traditional method of direct diagonalisation approach [51]. The SMMC is successful in the description of the Gamow-Teller strength distribution and has been applied to calculate the $2\nu\beta\beta$ decay matrix element of ^{48}Ca in complete pf-shell and ^{78}Ge in the upper fpg-shell. The predictive power of the SMMC can be established only after further studies.

Quasi-particle random phase approximation (QRPA)

The most extensively used approach for double beta decay matrix element calculations is the Quasiparticle Random Phase Approximation (QRPA). Basis of the QRPA method is a general Bogoluibov transformation and the basic idea is to re-express the particle degrees of freedom in terms of quasiparticle degrees of freedom. The calculation then consists of three main steps. The first one calculates the single particle energies and wave functions in a single particle potential. The dominant part of the remaining residual interaction between nucleons is the pairing force, which smears out the Fermi energy. It is taken into account in the second step of the calculation in the BCS formalism. The third step of the QRPA calculation is then to perform an RPA calculation to account for particle-hole and particle-particle excitations, which are assumed to have arbitrary phases. The advantage of QRPA calculations compared to more conventional shell model calculations are their abilities to treat large model spaces, typically consisting of two major oscillator shells. However, QRPA calculations contain several approximations such as the quasiboson approximation, etc. In double beta decay the initial nucleus decays into the final nucleus through the virtual excitation of all possible states of the intermediate nucleus. In the QRPA taking advantage of the harmonic approximation a large number of intermediate states can be constructed in a small model space. The quenching of $M_{GT}^{2\nu}$ is achieved in the QRPA through the proper inclusion of ground state correlations through the particle-particle interaction in the S=1, T=0 channel and the calculated half-lives almost agree with all experimental data by tuning the value of the strength of the particle-particle force to the available β^+ and β^- Gamow-Teller strengths. The pn-QRPA frequently overestimates the ground state correlations as a result of an increase in the strength of attractive proton-neutron interaction leading to the collapse of QRPA solutions. The physical value of this force is usually close to the point at which the QRPA solutions collapse. The origin of this unfortunate behaviour is attributed to the quasi-boson approximation (QBA) of the QRPA operators violating the Pauli exclusion principle. With increase in ground state correlations, the violation of the exclusion principle is also increased. To avoid this difficulty, renormalised QRPA (RQRPA) has been proposed which approximately takes into account the Pauli exclusion principle. The QRPA calculations including renormalisations do not fulfill the Ikeda sum rule by 30% and the calculated matrix elements are about smaller than earlier calculations full filling the sum rule by 40% properly [52] and [53]. In the QRPA approach a larger uncertainty may occur in deformed nuclei as deformation is usually ignored. This has been shown for^{150}Nd using QRPA and pseudo SU(3) model [54]. The nuclear transition matrix elements have been calculated for all double beta decay emitters and it has been shown that typical uncertainties of calculated $0\nu\beta\beta$ decay rates due to the limited knowledge of particle-particle force is of the order of a factor of two [53].

Projected Hartree-Fock-Bogoliubov (PHFB) model

The mass region 100 to 150 offers an example of shape transition i.e. the sudden onset of deformation at neutron number N=90. Nuclei range from spherical to well deformed, with large static quadrupole moments. Thus, it is expected that pairing and deformation degrees of freedom will play some crucial role in the structure of even-even $^{124-132}$Xe, $^{124-130}$Te and ^{130,132}Ba nuclei. The $\beta\beta$ decay can be studied in the same framework as many other nuclear properties and decays. The effect of deformation on the distribution of the Gamow-Teller and β-decay properties has been studied using a quasiparticle Tamm-Dancoff approximation (TDA) based on deformed Hartree-Fock (DHF) calculations with Skyrme interactions [56], a deformed self consistent HF+RPA method with Skyrme type interactions [57]. The comparison of the experimental GT strength distribution $B(GT)$ from its decay with the results of QRPA calculations was employed as a novel method of deducing the deformation of the $N = Z$ nucleus ^{76}Sr [58]. A deformed QRPA formalism to describe simultaneously the energy distributions of the single beta GT strength and the 2ν $\beta\beta$ decay matrix elements, using deformed Woods-Saxon potentials and deformed Skyrme Hartree-Fock mean fields was developed [59]. In all these works calculations are performed in the intrinsic basis, where angular momentum is not a good quantum number. The Projected Hartree-Fock-Bogoliubov (PHFB) model offers, in this sense, a competitive alternative. On one hand, the PHFB model restores the rotational symmetry, providing very reliable wave functions for the parent and daughter $\beta\beta$ decay emitters. On the other hand, in its present version the PHFB model is unable to provide information about the structure of the intermediate odd-odd nuclei, and, in particular, on the single β decay rates and the distribution of Gamow-Teller strength. Notwithstanding, the PHFB model has been successfully applied to the 2ν $\beta^-\beta^-$ decay of many emitters in the mass region $A \sim 100$, where it was possible to describe, in the same context, the lowest excited states of the parent and daughter nuclei, as well as their electromagnetic transition strengths, and to reproduce their measured $\beta\beta$ decay rates on the other [60].

A quantitative description of the 2ν $\beta^-\beta^-$ decay [60] and 2ν e^+DBD modes [61] of nuclei in the mass region $A \sim 100$ for the $0^+ \rightarrow 0^+$ transition has been obtained. The same formalism allowed the analysis of other observed nuclear properties, including the yrast spectra, the reduced transition probabilities $B(E2\!: 0^+ \rightarrow 2^+)$, the static quadrupole moments $Q(2^+)$ and the g -factors $g(2^+)$ of both parent and daughter nuclei. The study was performed using the PHFB model. Its application, in conjunction with the summation method, has motivated the present study of the 2ν $\beta^-\beta^-$ decay of ^{128,130} Te, ^{150}Nd and 2ν e^+DBD modes of ^{124,126}Xe and ^{130,132}Ba isotopes. The 2ν e^+DBD modes of ^{124,126}Xe and ^{130,132}Ba isotopes were studied in the earlier work of Shukla *et al.* [62]. In the present work, the HFB wave functions are generated with higher accuracy and in these nuclei, it is observed that the results are very much dependent on the later. The theoretical formalism to calculate the half-lives of 2ν $\beta\beta$ modes has been given in a number of reviews [63, 49]. Details of the PHFB study were presented in previous publications, for 2ν $\beta^-\beta^-$ decay [60] and 2ν e^+DBD modes [61] of nuclei in the mass region $A \sim 100$. Details of the mathematical expressions used to calculate the spectroscopic properties of nuclei in the PHFB model have been given by Dixit *et al.* [64]. The role of deformation on NTMEs $M_{2\nu}$ is also studied for these nuclei.

Other Methods

Although there are many other approaches that are being tried for evaluation of DBD nuclear matrix elements. Two of our interest in present context are $SU(4)_{\sigma\tau}$ and Pseudo SU(3). Wigner's supermultiplet symmetry $SU(4)_{\sigma\tau}$ has been used to understand the physical mechanism responsible for the quenching of [34]. Pseudo SU(3) is one of the few approaches which have been considered recently [54, 55] to take care of deformation in the calculation of double beta decay matrix elements. In the pseudo SU(3) shell-model coupling scheme [54] the normal-parity orbitals are identified with orbitals of a harmonic oscillator of one quantum less and these normal-parity orbitals are transformed into the pseudo space through an analytic expression. The double beta decay, when described in the pseudo SU(3) scheme, is strongly dependent on the occupation numbers for protons and neutrons in the normal and abnormal parity states which are determined by filling the Nilsson level. Using this approach, it has been confirmed that deformed nuclei have smaller nuclear matrix elements than the spherical nuclei.

Thus there are mainly two types of models usually employed to calculate 2ν and $0\nu\beta\beta$ decay nuclear transition matrix elements. One is based on the shell model and the second uses random phase approximation and extensions their of. Both the shell model and the QRPA model have succeeded in describing the DBD nuclear matrix elements for many cases, still they having their own limitations. PHFB model is proving to be one of the competitive alternative models particularly for deformed nuclei.

We see from our discussions on experimental and theoretical aspects of e^+DBD that the results from experiments are getting quite closer to the theoretical predictions. In particular the β^+EC is one of the best possible modes that can be achieved for some of the nuclei like ^{106}Cd and ^{130}Ba in near future in the ongoing/planned 2ν and possibly for 0νDBD experiments too. ^{106}Cd could be a stronger case as the active detector for ^{116}Cd is already proving to be successful[35].

4. Conclusion

Positron double beta decay is going to play important role once neutrinoless double beta decay is detected. There are some difficulties like low abundance of these nuclei in comparison to e^-DBD cases, but there is an experimental advantage in form of getting the additional signal of gamma rays in coincidence which can be important for identification of such rare processes. Another important reason for having at least one 0ν β^+EC experiment as pointed out earlier in the introduction is its necessity for distinguishing the process beyond standard model that could be responsible for the 0ν DBD once this mode is observed immaterial of e^-DBD or e^+DBD experiment in near future. Out of two experiments at least one has to be e^+DBD.

Acknowledgement: It is my pleasure to thank my colleagues and collaborators at IIT Kharagpur, Lucknow University, Gran Sasso and Ukrain. In particular I am greatful to P. K. Rath, R. Bernabei, V.I. Tretyak and F.A. Danevich for discussions at various stages. Help under Indo-Italian POC grant no. INT/Italy/POC2005-7 Ph.T-4)/2005 from DST India and MAE Italy made it possible to have timely closer interaction with LNGS experiments through R. Bernabei, the spokesperson of DAMA. Financial help from DST India and MAE Italy is greatfully acknowledged.

References

[1] A. M. Bakalyarov et al., Phys. Part. Nucl. Lett. 2 (2005) 77, hep-ex/0309016.

[2] I. Abt et al. [GERDA Collaboration], Nucl. Instr. and Meth. A 570/3 (2007) 479.arxiv:nucl-ex/0701004.

[3] C. Arnaboldi, et al., Phys. Rev. Lett. 95 (2005) 142501, hep-ex/0501034.

[4] L. Bardelli, et al., Nucl.Instrum.Meth. A569 (2006) 743-753,arXiv:nucl-ex/0608004.

[5] G. Audi and A.H. Wapstra, Nuclear Physics A 595, 409(1995).

[6] H.J. Kim et al., KEK Proc. 6 (2003) 205.

[7] R. Cerulli et al., Nucl. Instrum. Meth. Phys. Res. A 525 (2004) 535.

[8] K. Zuber, Eur. Phys. J. C 33 (2004) 817.

[9] P. Belli et al., Nucl. Phys. B 563 (1999) 97.

[10] P. Belli et al., Nucl. Instrum. Meth. Phys. Res. A 498 (2003) 352.

[11] M. Doi, T. Kotani and E. Takasugi, Prog. Theor. Phys. Suppl. 83, 1 (1985).

[12] M. Hirsch, K. Muto, T. Oda and H. V. Klapdor-Kleingrothaus, Z. Phys. A 347, 151 (1994).

[13] H.V. Klapdor-Kleingrothaus et al., Phys. Rev. D 73 (2006) 013010.

[14] V.I. Tretyak, Yu.G. Zdesenko, At. Data Nucl. Data Tables 61 (1995) 43.

[15] V.I. Tretyak, Yu.G. Zdesenko, At. Data Nucl. Data Tables 80 (2002) 83.

[16] J.H. Fremlin, M.C. Walters, Proc. Phys. Soc. A 65 (1952) 911.

[17] A. Berthelot et al., Comp. Rend. 236 (1953) 1769.

[18] E.B. Norman, Phys. Rev. C 31 (1985) 1937.

[19] H. Kiel, D. Munstermann, K. Zuber, Nucl. Phys. A 723 (2003) 499.

[20] F.A. Danevich et al., Nucl. Instrum. Meth. Phys. Res. A 544 (2005) 553.

[21] P. Belli et al., Preprint ROMA2F/2007/13July2007.

[22] T. Bloxham, et al., arXive: [nucl-exp] 07072756v.

[23] A.Sh. Georgadze et al., Phys. At. Nucl. 58 (1995) 1093.

[24] A.S. Barabash et al., Nucl. Phys. A 604 (1996) 115.

[25] F.A. Danevich et al., Z. Phys. A 355 (1996) 433.

[26] F.A. Danevich et al., Phys. At. Nucl. 59 (1996) 1.

[27] I. Stekl et al., Czech. J. Phys. 56 (2006) 505.

[28] A.S. Barabash, R.R. Saakyan, Phys. At. Nucl. 59 (1996) 179.

[29] A.P. Meshik et al., Phys. Rev. C 64 (2001) 035205.

[30] I. Bikit et al., Appl. Radiat. Isot. 46 (1995) 455.

[31] P. Belli et al., Astropart. Phys. 10 (1999) 115.

[32] R. Bernabei, Double Beta Decay meeting, Gran Sasso National Laboratory, November 2002.

[33] A. Bolozdynya et al., Nuclear Science Symposium, 1996. Conference Record, 1996 IEEE

[34] O. A. Rumyantsev and M. G. Urin, Phys. Lett. B 443 (1998) 51.

[35] F. A. Danevich, et al. Phys. Rev. C 68 (2003) 035501.

[36] S. Stoica and H.V. Klapdor-Kleingrothaus, Eur. Phys. J. A 17 (2003) 529.

[37] A. S. Barabash, et al. Nucl. Phys. A 604 (1996) 115.

[38] J. Suhonen and O. Civitarese, Phys. Lett. B 497 (2001) 221.

[39] E.B. Norman, M.A. De Faccio, Phys. Lett. B 148 (1984) 31.

[40] R. G. Winter, Phys. Rev. 99 (1955) 88.

[41] A. Staudt, K. Muto and H. V. Klapdor- Kleingrothaus, Phys. Lett. B 268 (1991) 312.

[42] J. Toivanen and J. Suhonen, Phys. Rev. C 55 (1997) 2314.

[43] J. Suhonen, Phys. Rev. C 48 (1993) 574.

[44] O. Civitarese and J. Suhonen, Phys. Rev. C 58 (1998) 1535.

[45] A. S. Barbash et al, Phys. Lett. B 223 (1989) 273.

[46] M. Aunola and J. Suhonen, Nucl. Phys. A 602 (1996) 133.

[47] Ju M. Gavriljuk et al, Phys. At. Nucl. 61 (1998) 1287.

[48] P. B. Radha et al., arXiv: nucl-th/9510052; Phys. Rev. Lett. 76, 2642 (1996).

[49] J. Suhonen and O. Civitarese, Phys. Rep. 300 (1998) 123.

[50] E. Caurier, F. Nowacki, A. Poves and J. Retamosa, Phys. Rev. Lett. 77, 1954 (1996).

[51] S. E. Koonin, D. J. Dean and K. Langanke, Phys. Rep. 278, 1(1997).

[52] K. Muto, E. Bender, H. V. Klapdor-Kleingrothaus, Z. Phys. A 334, 177 (1989).

[53] A. Staudt, K. Muto, H. V. Klapdor-Kleingrothaus, Europhys. Lett. 13, 31 (1990).

[54] J. G. Hirsch, O. Castaños, P. O. Hess and O. Civitarese, Phys. Rev. C 51 (1995) 2252.

[55] V. E. Ceron and J. G. Hirsch, Phys. Lett.B 471 (1999) 1.

[56] F. Frisk, I. Hamamoto, X.Z. Zhang, Phys. Rev. C **52**, 2468 (1995).

[57] P. Sarriguren, E. Moya de Guerra, A. Escuderos, A.C. Carrizo, Nucl. Phys. A **635**, 55 (1998); P. Sarriguren, E. Moya de Guerra, A. Escuderos, Nucl. Phys. A **658** (1999) 13; *ibid* Nucl. Phys. A **691**, 631 (2001).

[58] E. Nácher *et al.*, Phys. Rev. Lett. **92**, 232501(2004).

[59] R. Álvarez-Rodr´guez, P. Sarriguren, E. Moya de Guerra, L. Pacearescu, A. Faessler, F. Simkovic, Phys. Rev. C **70**, 064309 (2004).

[60] R. Chandra, J. Singh, P.K. Rath, P.K. Raina, J.G. Hirsch, Eur. Phys. J. A **23**, 223 (2005)

[61] P.K. Raina, A. Shukla, S. Singh, P.K. Rath, J.G. Hirsch, Eur. Phys. J. A **28**, 27 (2006); A. Shukla, P.K. Raina, R. Chandra, P.K. Rath, J.G. Hirsch, Eur. Phys. J. A **23**, 235 (2005).

[62] A. Shukla, P.K. Raina, P.K. Rath, J.Phys. G: Nucl. Part. Phys. **34**, 549 (2007).

[63] M. Doi, T. Kotani, Prog. Theor. Phy. **87**, 1207 (1992).

[64] B.M. Dixit, P.K. Rath, P.K. Raina, Phys. Rev. C **65**, 034311 (2002), Phys. Rev. C **67**, 059901(E) (2003).

NEUTRINOLESS DOUBLE BETA DECAY
Editors: V.K.B. Kota and U. Sarkar

POSITRON DOUBLE BETA DECAY IN CDZNTE MATERIALS

Manoj K. Sharan[1]

Saha Institute of Nuclear Physics, Kolkata - 700 064, India

Abstract

Investigation of positron double decay experimetns using CdZnTe semiconductor detector is being looked into. Background studies of CdZnTe crystals, obtained from CEL, New Delhi has been carried out. Simulation using EGS4 has been setup for these materials and are being carried out.

1. Introduction

In recent years there has been considerable interest in the study of $\beta^-\beta^-$ decay modes[1, 2, 3, 4, 5]. Howevery it is necessary to explore other unconventional processes like transition into excited states and double positron decay or double electron capture. In principle, such studies give same information as electron decay modes. Theoretically such studies also have possiblities of giving information about dominance of Majorana neutrino mass or the right handed admixures in weak interactions[6]. Also, the experimental sensitivity of $\beta^-\beta^-$ decay modes gets limited because of the presence of electron bakcground. From the experimental point of view positron decay modes have the possiblity of serving as a convenient tool since they are relatively easier to be separated from the background contaminations.

[1] **E-mail:Manoj.Sharan@saha.ac.in**

Table 1. Decay modes, Q-values and natural abundance of isotopes of positive beta decay

Isotope	Decay mode	Q-value (keV)	Nat. abund. (%)
^{58}Ni	β^+/EC, EC/EC	1926	68.08
^{64}Zn	β^+/EC, EC/EC	1096	48.63
^{78}Kr	$\beta^+\beta^+$, β^+/EC, EC/EC	2866	0.35
^{92}Mo	β^+/EC, EC/EC	1649	14.84
^{96}Ru	$\beta^+\beta^+$, β^+/EC, EC/EC	2719	5.54
^{106}Cd	$\beta^+\beta^+$, β^+/EC, EC/EC	2771	1.25
^{120}Te	β^+/EC, EC/EC	1698	.09
^{124}Xe	$\beta^+\beta^+$, β^+/EC, EC/EC	2865	.09
^{130}Ba	$\beta^+\beta^+$, β^+/EC, EC/EC	2611	.11
^{136}Ce	$\beta^+\beta^+$, β^+/EC, EC/EC	2400	14.84

2. Detector Considerations for $2\beta^+$ Decay Experiment

Of the known 35 known isotopes able to undergo double beta decay, the istopes of positve beta decay, their decay modes, Q-values and natural abundance is listed in Table 1. Positron double beta decay experiments suffer from poor net abundance of $2\beta^+$ decay candidates and also lower Q values.

Because of these reasons, current level of experimental sensitivity offers half-life limits in the range of $10^{18} - 10^{20}$ years[9]. The half lives for $2\beta^+$ decay searches are quite modest in comarison to the already achieved $2\beta^-$ decay modes. However, the present thoretical expectations for some of the $2\beta^+$ istopes have found to be quite favourable[6].

With improved sensitivity in detection systems, it is expected that the positron double beta decay modes will be obseved in the near future.

The important considerations for 2μ positron doublebeta decay experiments are -

• Use of abundant isotopes

• Good efficiency of detector systems or by designing an experimental setup where the detector is also the source.

• A large mass detector

• A good energy resolution

• A large Q value as it increases the phase space for double beta decay. Also the background rate generally decreases with increasing energy and the line energy resolution improves.

• Effective background suppressing techniques.

3. CdTe (CdZnTe) Crystals

In general, semiconductor detectors offer very good energy resolution and are produced in a clean environment. Further, semiconductor detectors made of CdZnTe crys-

Table 2. Double beta Isotopes present in CZT

Isotope	Nat. Abundance (%)	Q (keV)	Decay Mode
^{70}Zn	0.62	1001	$\beta^-\beta^-$
^{114}Cd	28.7	534	$\beta^-\beta^-$
^{116}Cd	7.5	2805	$\beta^-\beta^-$
^{128}Te	31.7	868	$\beta^-\beta^-$
^{130}Te	33.8	2529	$\beta^-\beta^-$
^{64}Zn	48.6	1096	β^+/EC
^{106}Cd	1.21	2771	$\beta^+\beta^+$
^{108}Cd	0.9	231	EC/EC
^{120}Te	0.1	1722	β^+/EC

tals can be operated at room temperature. These characteristics make it very useful for designing experiments particularly in an underground laboratory.

Cadmium Zinc Telluride (CZT) semiconductors offer very good potential in search for neutrinoless double beta decay[7, 8]. The material contains 5 isotopes which can undergo double beta decay (out of the total no. of 35 known isotopes). Table 2 shows the percentage, natural abundance in the crystals, the Q value of the decay and the decay mode. Four of the isotopes can also undergo double positron decay, Due to energy constraints only ^{106}Cd can emit two positrons, whilst the other isotopes decay via single electron capture of a K-shell electron (β^+/EC) and double electron capture (EC/EC) modes. In addition, the atoms can be left in excited states after the decay, subsequently emitting gamma rays to return the atom to the ground state.

In CZT material, ^{130}Te has the lowest Q value at 2.529 MeV but has the advantage of having the highest natural abundance at 33.8%. ^{116}Cd is also an interesting candidate as it has a reasonably good abundance at 7.5% and its Q value (2.805 MeV) is greater than the highest energy γ-ray emitted in the decay of natural thorium (2.614 MeV) and thus should help in eliminating background from surrounding rocks. For the study of $2\beta^+$ decay processes, ^{106}Cd is considered to be a very good candidate because of its high Q value (2.771 MeV) and favourable theoretical estimates for half lives[6, 10]. It is hoped that the development of detectors of enriched isotopes of CZT would help in studying different aspects of double beta decay.

CdZnTe crystals have also considerable industrial and scientific applications and lot of developments are taking place in detector fabrication[11, 12].

A facility for production of Cd-Zn-Te single crystal exists at Central Electronics Limited (CEL), Sahibabad (CSIR laboratory), India[13]. These crystals are produced by Asymmetric Bridgeman tecnique using pure materials of Cd, Zn, Te (all are of purity 99.9999999%

The processes for production include synthesis, crystal growth and annealing of ingots. After the crystal growth, wafers are sliced to desired lengths and lapping and polishing is done. The subtrates produced have electrical resistivity of $10^5 - 10^8$ ohm-cm.They also have excellent properties like infra-red transmission $> 55\%$.

4. Background Studies

The CdTe crysal (1 mm) in thickness were obtained from CEL and the sample was checked for any contamination. For this purpose, a HPGe clover detector set up was used at SINP[14]. The HPGe detector was placed inside a cubicle box with graded shielding with copper, Cd and lead to reduce the background. The sample(CdTe) was placed in front of the detector. Gamma-ray spectrum (50 keV to 3 MeV) was recorded in a PC based MCA, whose long term stability was checked. The detector was calibrated using ^{60}Co, ^{133}Ba and ^{152}Eu source. Spectrum was recorded with and without the sample for 24 hours each. Stability of the detectors and the setup was checked at regular intervals. We did not find gamma-ray line from any contamination in the sample. We are further planning to make CdTe Schottky diode detectors from the crystals supplied by CFL.

5. Test of CdTe detector

We are in the process of testing a CdTe detector whose area is 1.83X1.83 mm^2 and whose volume is 8 mm^3. These detectors were obtained from Eurorad by the Plasma Physics Laboratory of SINP. The operating bias of detector is 50 V and leakage current is 5 nA. These detectors can be operated at room temperature and thus any cryostat is not necessary. We have tested the pre-amplifier and checked with pulsar. The spectrum had large pick up signal which has been eliminated. Currently we are working on reducing the background pickup signal. We intend to check the detector with standard sources like ^{137}Cs and ^{57}Co. The resolution, background and long term stability of the detector will be tested.

6. Simulation (EGS4)

EGSnrc system is a package for the Monte Carlo (MC) simulation of coupled electron-photon transport. The current energy range of applicability is considered to be 1 keV - 10 GeV. EGSnrc is an extended and improved version of the EGS4 package originally developed at SLAC[15].

We have set the code on Red Hat Linux for CdTe detectors. User routines have been written for different detector geometries and response of the detector has been simulated. In figure 1, response of a CdTe detector for source ^{57}Co is shown. In figure 2, the response of the detector for source ^{137}Cs is shown. The spectral shapes match reasonably well with experimentalresults. In figure 3, we have shown the response of ^{40}K which is randomly present as bakcground.

More detailed simulations for different detector geometries are in progress.

7. Plans and Summary

We plan to characterize the existing CdTE detector. We also plan to obtain CdTe crystals from CEL and make CdTe Schottky detectors. These need to be checked with simulations which are being carried out. We also intend to explore the possibility of making large scale detectors and check their stability. Background studies at the site are also planned.

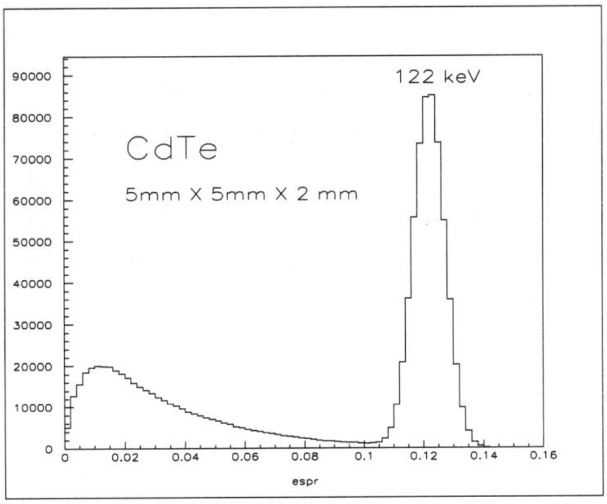

Figure 1. Simulated spectrum of CdTe detector with source ^{57}Co

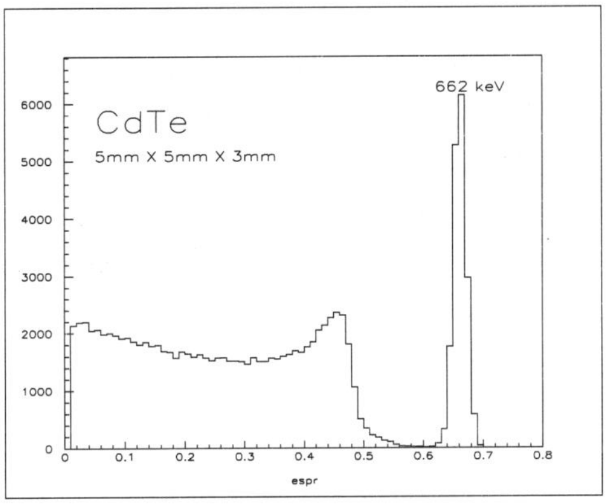

Figure 2. Simulated spectrum of CdTe detector with source ^{137}Cs

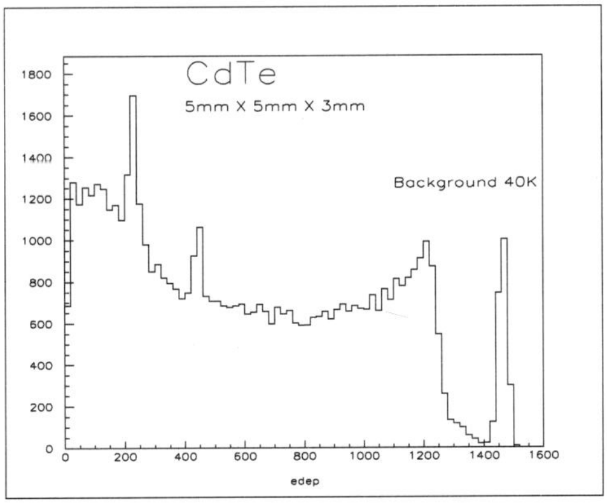

Figure 3. Simulated background spectrum of CdTe detector assuming 40K to be uniformly distributed inside the detector.

Acknowledgement

The author wishes to thank Prof. P.K. Raina for his help in all aspects of this new field. We thank Dr. Rakesh Kumar Sinha of CEL for providing us with CdTe crystals for background tests. We are thankful to Dr. R. Narayanan, P. Bhattacharya and Prof. S. Iyengar for providing us with CdTe detectors for carrying out tests. We thank Prof. S. Bhattacharya, Prof. A. Goswami and R. Rout for their help in carrying out the background studies.

References

[1] C. Dörr, H.V. Klapdor-Kleingrothaus, *Sixty years of double beta decay*, World Scientific, Singapore, (2001)

[2] H. Ejiri, Prog. in Part. and Nucl. Phys. **40**, (1998) 307

[3] H. Ejiri, Prog. in Part. and Nucl. Phys. **48**, (2002) 185

[4] S.R. Eliott and P. Vogel, Ann. Rev. Nucl. Part. Sci. **52**, (2002) 115

[5] R.D. McKeown, P. Vogell, Phys. Rep. **394**, (2004) 316

[6] M. Hirsch, M. Muto, T. Oda, H.V. Klapdor-Kleingrothaus, Z. Phys. **A 347**, (1994) 51.

[7] K. Zuber, Phy. Lett. B **519**, (2001) 1

[8] H. Kiel, D. Münstermann, K. Zuber, Nucl. Phys A **723**, (2003) 499

[9] V.I. Tretyak, Yu.G.Zdesenko, At. Data Nucl. Data Tables **61**, (1995) 43.

[10] J. Suhonen, Phys. Rev. **C 48**, (1993) 574.

[11] Tadayuki Takahashi and Shin Watanabe, IEEE Transactions on Nuclear Science, **48** (2001) 950

[12] T. Takahashi et al, Nucl. Instum. Methods. A **541**, (2005) 332

[13] R.K. Sinha et al, DAE Symp. in High Energy Physics

[14] Sudeb Bhattacharya et al, DAE Symp. in High Energy Physics.

[15] I. Kawrakow, E. Mainegra-Hing and D.W.O. Rogers, NRCC Report *PIRS-877* (2006)

NEUTRINOLESS DOUBLE BETA DECAY
Editors: V.K.B. Kota and U. Sarkar

RESULTS ON DARK MATTER AND $\beta\beta$ DECAY MODES BY DAMA AT GRAN SASSO

R. Bernabei[1]

Dip. di Fisica, Università di Roma "Tor Vergata" and INFN-Roma Tor Vergata, 00133 Roma, Italy

Abstract

DAMA is an observatory for rare processes and it is operative deep underground at the Gran Sasso National Laboratory of the I.N.F.N. (LNGS). Here some arguments will be presented on the investigation on dark matter particles by annual modulation signature and on some of the realized double beta decay searches.

1. Introduction

DAMA is an observatory for rare processes and it is operative deep underground at the Gran Sasso National Laboratory of the I.N.F.N. (LNGS). The main experimental set-ups are: i) DAMA/NaI ($\simeq$ 100 kg of highly radiopure NaI(Tl)), which completed its data taking on July 2002 [1, 2, 3, 4, 5, 6, 7, 8, 9, 10, 11, 12, 13, 14, 15, 16, 17, 18,

[1] **e-mail:rita.bernabei@roma2.infn.it**

19, 20, 21, 22, 23]; ii) DAMA/LXe ($\simeq$ 6.5 kg liquid Kr-free Xenon enriched either in ^{129}Xe or in ^{136}Xe) [24, 25, 26, 27, 28, 29, 30, 31, 32, 33, 34, 35, 36, 37, 38]; iii) DAMA/R&D, devoted to tests on prototypes and to small scale experiments, mainly on the investigations of double beta decay modes in various isotopes [39, 40, 41, 42, 43, 44, 45, 46, 47, 48]; iv) the new second generation DAMA/LIBRA set-up ($\simeq$ 250 kg highly radiopure NaI(Tl)) in operation since March 2003 [49]; v) the low background DAMA/Ge detector mainly devoted to sample measurements; in some measurements the low-background Germanium detectors of the LNGS facility are also used [50, 51]. Moreover, a third generation R&D is in progress towards a possible 1 ton set-up, DAMA proposed in 1996.

2. The Investigation on Dark Matter Particles by Annual Modulation Signature

2.1. The DAMA/NaI Model Independent Result

The highly radio-pure DAMA/NaI set-up [3, 7, 11, 12] has been a pioneer Dark Matter experiment of suitable exposed mass, sensitivity and stability of the running conditions, taking data at LNGS over seven annual cycles. The main aim of DAMA/NaI has been the investigation of the presence of Dark Matter (DM) particles in the galactic halo by exploiting the model independent annual modulation signature, originally suggested by [52, 53] in the middle of 80's. The data taking has been completed in July 2002 and still results are produced by various kinds of studies. The final model independent result on Dark Matter particles by DAMA/NaI has been published in [11, 12], while some of the many possible corollary model dependent investigations have been published in [11, 12, 13, 14]; other corollary quests are available in literature, such as e.g. refs. [54, 55, 56, 57, 58, 59, 60]. Moreover, many other scenarios can be considered as well and some of them are already under analysis. In addition, profiting by its low-background features and by the high collected exposure, several results have been achieved both on Dark Matter particle investigations with other approaches and on several other rare processes [1, 15, 16, 17, 18, 19, 20, 21, 22, 23].

We just remind that the annual modulation signature is based on the yearly Earth motion around the Sun, which is moving in the galactic halo. This motion induces a "wind" of Dark Matter particles on the Earth, whose intensity – measured by the experimental setup deep underground – varies along the year. This intensity should satisfy the following peculiarities: i) it varies as a cosine function; ii) with 1 year period; iii) with maximum around roughly 2 June; iv) with amplitude not exceeding 7% in many of the possible scenarios; but some scenarios exist which can expect a larger value. Moreover, the variation should be: v) present only in the low energy region where Dark Matter particles can induce the signal; vi) present only in events where energy is deposited only in one of the several scintillators of the setup since the probability of multiple interactions is negligible. The highly radiopure NaI(Tl) is particularly suitable for this kind of experiment since it is sensitive to Dark Matter particles of various natures, both of low (through Na target) and high mass (through I target) and with various type of interactions; moreover, it also offers many technical-practical advantages. In the DAMA/NaI experiment advanced techniques for low background experiments have been exploited and further developed; a continuous monitoring and control of all the running conditions was also assured.

This signature, when analysing the measured counting rate in a suitable large mass low background NaI(Tl), offers the possibility to point out the presence of Dark Matter particles in the galactic halo in a model independent way.

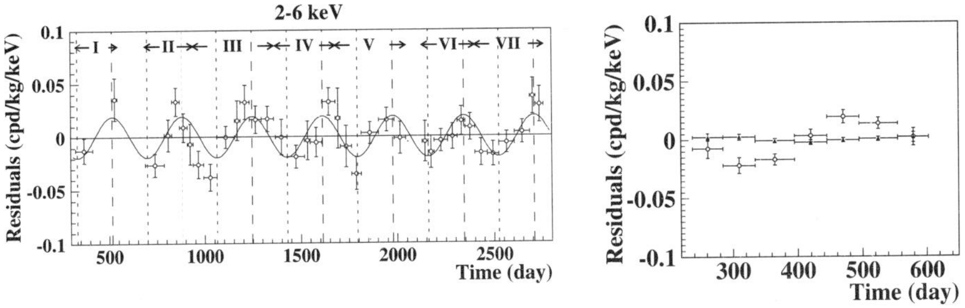

Figure 1. *Left:* experimental residual rate for *single-hit* events in the cumulative (2–6) keV energy interval as a function of the time over 7 annual cycles, end of data taking July 2002. The experimental points present the errors as vertical bars and the associated time bin width as horizontal bars. The superimposed curve represents the cosinusoidal function behaviour expected for a Dark Matter particle signal with a period equal to 1 year and phase exactly at 2nd June; the modulation amplitude has been obtained by best fit [11, 12]. *Right:* experimental residual rates over seven annual cycles for *single-hit* events (open circles) – class of events to which DM events belong – and over the last two annual cycles for *multiple-hits* events (filled triangles) – class of events to which DM events do not belong – in the (2–6) keV cumulative energy interval. They have been obtained by considering for each class of events the data as collected in a single annual cycle and using in both cases the same identical hardware and the same identical software procedures. The initial time is taken on August 7th. This latter result offers an additional strong support for the presence of DM particles in the galactic halo further excluding any side effect either from hardware, from software procedures or from background.

In particular, the DAMA/NaI set-up has pointed out the presence of an annual modulation in the *single-hit* residual rate in the lowest energy interval (2 – 6) keV (see Fig. 1); the observed effect satisfies many peculiarities of the DM particle induced signature. This gives a 6.3 σ C.L. evidence over seven annual cycles (total exposure: 107731 kg $\times$ day) for the presence of DM particles in our Galaxy (see refs. [2, 4, 5, 6, 7, 8, 9, 10, 11, 12, 13, 14]; moreover, further works on the corollary model dependent investigations are in progress). Neither systematic effects nor side reactions able to account for the measured modulation amplitude and to contemporaneously satisfy all the peculiarities of the signature have been found or suggested by anyone.

This result is model-independent. It represents the first experimental evidence of the presence of DM particles in the galactic halo independently on their composition and nature. At present, apart from DAMA/LIBRA, no other experiment is sensitive, for mass, radiopurity and control of the stability, to such a model independent signature and, in particular, no other activity is available whose result can directly be compared in a model independent way with the one of DAMA/NaI.

2.2. Some of the Possible Corollary Quests for the DM Candidate Particles

Several corollary investigations have been pursued on the nature of the DM candidate particles and on the astrophysical scenarios. These corollary investigations are instead model-dependent and, due to the poor knowledge on the astrophysical, nuclear and particle physics needed assumptions and on the related parameters, they have no general meaning as it is also the case of exclusion plots and of DM particle parameters evaluated in indirect detection experiments. Thus, these investigations should be handled in the most general way (see for discussions at some extent e.g. [11, 12, 13, 14] and in literature; other works are in progress).

Several (of the many possible) corollary quests for a candidate particle have been carried out on the class of DM candidate particles named WIMPs [11, 12]. Low and high WIMP mass candidates interacting with ordinary matter via the general case of mixed SI&SD coupling and the subcases of dominant SI coupling or of dominant SD coupling has been considered as well as the case of WIMPs with preferred SI inelastic scattering [11, 12]. In fact, the general solution for WIMPs is a 4-dimensional allowed volume in the space ($\xi\sigma_{SI}$,$\xi\sigma_{SD}$,θ,m_W) in the first case (being the two subcases given by slices of this volume) and a 3-dimensional allowed volume in the space ($\xi\sigma_p$,δ,m_W) for the last one [11, 12]. There, ξ is the fractional amount of local density of DM particles, σ_{SI} and σ_{SD} are the point-like SI and SD DM particle-nucleon cross sections and tgθ is the ratio between the effective SD coupling strengths of the DM particle with neutron and proton (θ is defined in the $[0,\pi)$ interval), σ_p is the cross section for WIMPs with preferred inelastic interactions and δ is the mass splitting of the two mass states of this candidate.

Accounting at some extent for the present poor knowledge of all the astrophysical, nuclear and particle Physics aspects the results have been given in terms of external contour obtained by the superposition of the allowed volumes/regions, that is they account for a large number of sets of best fit values. In particular, for the case of WIMP elastically scattering on nuclei the obtained best-fit mass ranges from GeV to about 1 TeV and the cross section spans over more than one order of magnitude when accounting at least at some extent for some of the present uncertainties on assumptions and parameters. For some details and results see refs. [11, 12]. Furthermore, Galaxy hierarchical formation theories [61], numerical simulations [62], the discovery of the Sagittarius Dwarf Elliptical Galaxy (SagDEG) in 1994 [63] and more recent investigations [64] suggest that the dark halo of the Milky Way can have a rich phenomenology containing non thermalized dark matter fluxes. Thus, recently the contribution of the SagDEG has been investigated by analyzing the consequences of its DM stream contribution to the galactic halo on the DAMA/NaI annual modulation data. For simplicity, in this analysis we have investigated these topics just considering some of the candidates belonging to the WIMP class and few of the many possible astrophysical, nuclear and particle Physics scenarios as already considered in refs. [11, 12] in order to allow a direct comparison. Some examples are given in Fig. 2, while results on further aspects and details can be found in ref. [14].

Other streams can potentially play more intriguing roles in the corollary investigations for whatever candidate particle with whatever approach and for comparisons; they will be investigated in near future, such as e.g. the Canis Major [64] and those arising from caustic halo models [65].

The DAMA/NaI allowed volumes/regions are well compatible with theoretical expectations for neutralino in various scenarios (see as an example [56]). It is worth to remind that the neutralino itself has both SI and SD coupling and that the basic theory has a very large number of "open" parameters. Also several other WIMP candidates fit

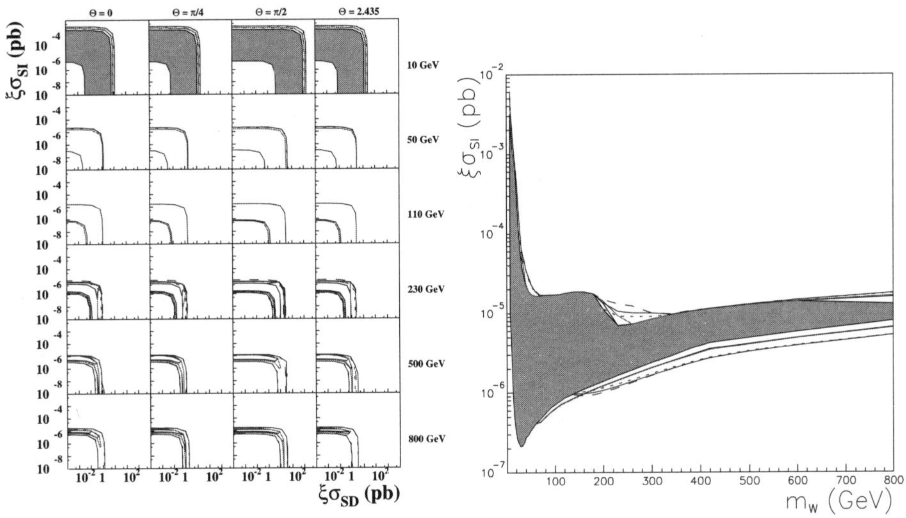

Figure 2. Left: filled areas are examples of slices, in the plane $\xi\sigma_{SI}$ vs $\xi\sigma_{SD}$, of the 4-dimensional allowed volume for DM particle with mixed SI&SD interaction for the model frameworks given in [11, 12]. Right: filled area is the region allowed in the $(\xi\sigma_{SI}, m_W)$ plane in the scenarios for pure SI coupling considered in [11, 12]. The areas enclosed by the lines are obtained by introducing the SagDEG contribution [14]. In all the cases inclusion of other existing uncertainties on parameters and models would further extend the regions and increase the sets of best fit values.

the DAMA/NaI model independent evidence as well as other existing candidates with a phenomenology similar, but not identical as for the WIMP cases e.g.: the mirror Dark Matter particles [59], the self-interacting dark matter particles [60], etc. etc. Moreover, several candidates not belonging to the WIMP class and having different interaction features, to which DAMA/NaI – on the contrary of others – is fully sensitive, are also available. In addition, in principle whatever particle with suitable characteristics, even not yet foreseen by theories, can be a good candidate as DM in the galactic halo.

As a relevant case, the DAMA/NaI annual modulation data have also been analysed in terms of a light ($\simeq$ keV mass) bosonic candidate, either with pseudoscalar or with scalar axion-like coupling; details are available in ref. [13]. For these candidates, the direct detection process is based on the total conversion in NaI(Tl) crystal of the mass of the absorbed bosonic particle into electromagnetic radiation. Thus, in these processes the target nuclei recoil is negligible and is not involved in the detection process. Thus, signals from these light bosonic DM candidates are detected in DAMA/NaI, but are lost in activities based on the selection/rejection procedures of the electromagnetic contribution to their counting rate (such as e.g. Cdms, Edelweiss, Cresst, Warp, etc.) [13]. Despite these particles are unstable, their lifetime can be of cosmological interest and offers valuable candidates for the DM signal, observed in DAMA/NaI. In particular, the result for the pseudoscalar case for the same astrophysical scenarios of ref. [11, 12] is shown in Fig. 3.

Obviously, the discussed corollary quests are not at all exhaustive of the many scenarios open at present level of knowledge for these and for other classes of candidates as well as for the astrophysical and nuclear/atomic aspects. Several other investigations

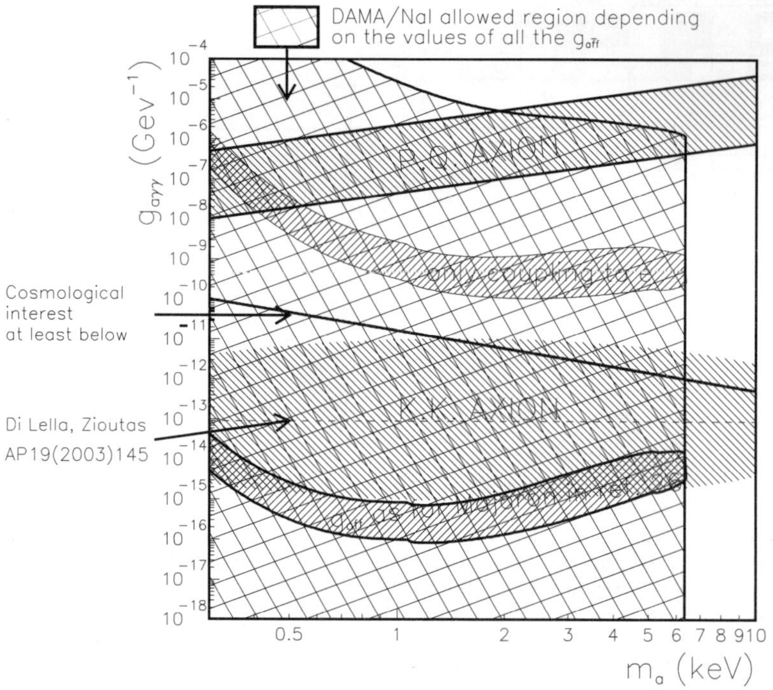

Figure 3. DAMA/NaI allowed region at 3σ C.L. in the plane $g_{a\gamma\gamma}$ vs m_a for a light pseudoscalar candidate (*crossed hatched region*). All the configurations in this region can be allowed depending on the values of all the $g_{a\bar{f}f}$. Examples of two of the many possible models: i) *upper black region*: coupling only to electrons; ii) *lower black region*: coupling (proportional to the m_a) through the (weak) isospin to quarks and leptons. For details see ref. [13]. The indicative region of the Kaluza-Klein (K.K.) pseudoscalar axion credited in ref. [66] from the analysis of indirect observations and the region of the DFSZ and KSVZ models (P.Q. axion) are shown as well. The solid line corresponds to *a* particle with lifetime equal to the age of the Universe; at least all the $g_{a\gamma\gamma}$'s below this line are of cosmological interest. However, in principle, it might be possible that the configurations above this line would also become of interest in case of some exotic mechanism of the *a* particle production. Thus, a pseudoscalar DM candidate can also account for the DAMA/NaI model independent result as well as the WIMP solutions already discussed elsewhere.

have been discussed in literature and other studies are in progress.

2.3. Few Words on Comparisons in the Field

No experiment is available – with the exception of the presently running DAMA/LIBRA – whose result can be directly compared in a model independent way with the experimental positive 6.3 σ C.L. evidence for the presence of particle Dark Matter in the galactic halo obtained by DAMA/NaI. Thus, claims for contradictions have intrinsically no scientific meaning.

In particular, as regards some claimed model-dependent comparisons presented so far we just mention – among the many existing arguments – that: i) the other experiments are insensitive to the annual modulation signature; ii) they release just a marginal exposure (orders of magnitude lower than the one by DAMA/NaI) after several/many years underground; iii) they exploit strong data selection and strong and often unsafe rejection techniques of their huge counting rate, becoming at the same time insensitive to several DM candidates; iv) they generally quote in an incorrect/partial/not updated way the DAMA/NaI result; v) they consider a single model fixing all the astrophysical, nuclear and particle Physics assumptions as well as all the theoretical and experimental parameters at a single questionable choice [2]. Thus, e.g. for the WIMP case they do not account for the existing uncertainties on the real coupling with ordinary matter, on the spin-dependent and spin-independent form factors and related parameters for each nucleus, on the spin factor used for each nucleus, on the real scaling laws for nuclear cross sections among different target materials; on the experimental and theoretical parameters, on the effect of different halo models and related parameters on the different target materials, etc. For example, large differences are expected in the counting rate among nuclei fully sensitive to the SD interaction (as ^{23}Na and ^{127}I) with the respect to nuclei largely insensitive to such a coupling (as e.g ^{nat}Ge, ^{nat}Si, ^{nat}Ar, ^{nat}Ca, ^{nat}W, ^{nat}O) and also when nuclei in principle all sensitive to this coupling but having different unpaired nucleon (e.g. neutron in case of the odd spin nuclei, such as ^{129}Xe, ^{131}Xe, ^{125}Te, ^{73}Ge, ^{29}Si, ^{183}W and proton in the ^{23}Na and ^{127}I). Moreover, in case the detection of the DM particles would involve electromagnetic signals (see, for example, the case of the light bosons discussed above, but also electromagnetic contribution in WIMP detection arising e.g. from known effect induced by recoiling nuclei), all the experiments, such as e.g. Cdms, Edelweiss, Cresst, Zeplin, Warp and their extensions, do lose the signal in their rejection procedures of the e.m. contribution to the counting rate. In addition, those experiments present many critical points e.g. both regarding the energy scale and the multiple selection procedures (related stabilities and efficiencies, systematics in the claimed rejection factors, ranging from 10^{-4} to 10^{-8}, evaluations, stabilities and monitoring of the spill-out factors, ...) on which their claimed sensitivities (in a "single" set of assumptions and parameters' values) are based.

For completeness, it is also worth to note that no results obtained with different target material can intrinsically be directly compared even for the same kind of coupling, although apparently all the presentations generally refer to cross section on the nucleon. In fact, this requires – among others – the knowledge of e.g. the real scaling laws, form factors, etc.; the situation is much worse than that in the field of double beta decay experiments when different isotopes are used.

[2] We note that the naive and partial "prescription" of ref. [67] on some aspects for a single particular WIMP case cannot be defined – on the contrary of what appears in some paper – as a "standard theoretical model". Such a paper summarized a single oversimplified approach adopted at that time. Its use as "unique" reference is obviously incorrect, since it did not account at all for the level of knowledge on all the involved astrophysics, nuclear and particle physics aspects and parameters, for the many possibilities open on the astrophysical, nuclear and particle physics aspects and for the different existing approaches as e.g. the annual modulation signature. It is also worth to note that – in addition – the figure they usually reported for comparison is not only for the reasons, discussed here and elsewhere, in the substance meaningless, but it is intrisencally wrong also because e.g.: while they do not consider the model dependent LEP limit in their exclusion plots, they do not remove it from their "choice" of DAMA/NaI allowed space. If done, this will show – as known since time (see literature) – that even under their single arbitrary and naive scenario (which in addition cannot be applied at all to the annual modulation data which requires – among others – time/energy correlation analysis), they are not discriminative at all.

Discussions at some extent are also reported e.g. in ref. [11, 12, 13, 14]; all those general comments also hold in the substance for more recent data releases.

As regards the indirect searches, a comparison would always require the calculation and the consideration of all the possible DM particle configurations in the given particle model, since it does not exist a biunivocal correspondence between the observables in the two kinds of experiments. However, the present positive hints provided by indirect searches are not in conflict with the DAMA/NaI result.

Finally, it is worth to note that – among the many corollary aspects still open – there is f.i. the possibility that the particle dark halo can have more than one component. The more sensitive NaI(Tl) experiments in progress and in preparation by DAMA will also have this aspect among the arguments to be further investigated.

2.4. The New DAMA/LIBRA and Beyond

In 1996 DAMA proposed to realize a ton set-up and a new R&D project for highly radiopure NaI(Tl) detectors was funded. As a consequence of the results of this second generation R&D, the new experimental set-up DAMA/LIBRA (Large sodium Iodide Bulk for RAre processes), $\simeq$250 kg highly radiopure NaI(Tl) crystal scintillators (matrix of twenty-five $\simeq$ 9.70 kg NaI(Tl) crystals), was funded and realised. This set-up has replaced the previous DAMA/NaI; the experimental site as well as many components of the installation itself have been implemented (environment, shield of the photomultipliers, wiring, High-Purity Nitrogen system, cooling water of air conditioner, electronics and data acquisition system, etc.). In particular, all the Copper parts have been chemically etched before their installation following a new devoted protocol and all the procedures performed during the dismounting of DAMA/NaI and the installation of DAMA/LIBRA detectors have been carried out in High-Purity Nitrogen atmosphere [49].

DAMA/LIBRA is continuously taking data since March 2003; for example, in March 2007, an exposure of about 1.5×10^5 kg $\times$ day has already been collected. As in the case of the DAMA/NaI experiment regular routine calibrations are performed, e.g. in 4 years of measurements about 4×10^7 events of sources for energy calibration and 2×10^6 events/keV for the evaluation of the acceptance window efficiency have been acquired.

A large work is going to be faced by the new DAMA/LIBRA, which will also further investigate with higher sensitivity several other rare processes. The first data release is foreseen not later than 2008.

A third generation R&D effort towards a possible NaI(Tl) ton set-up, we proposed in 1996, has been funded by I.N.F.N. and related works on selection of materials for detectors and photomultipliers are in progress [68].

3. The Searches for Double Beta Decay Modes

The various low background set-ups, realized by the DAMA collaboration, also allow to investigate several kinds of rare processes. In particular, several double beta decay modes have been investigated by using either the active or the passive source technique

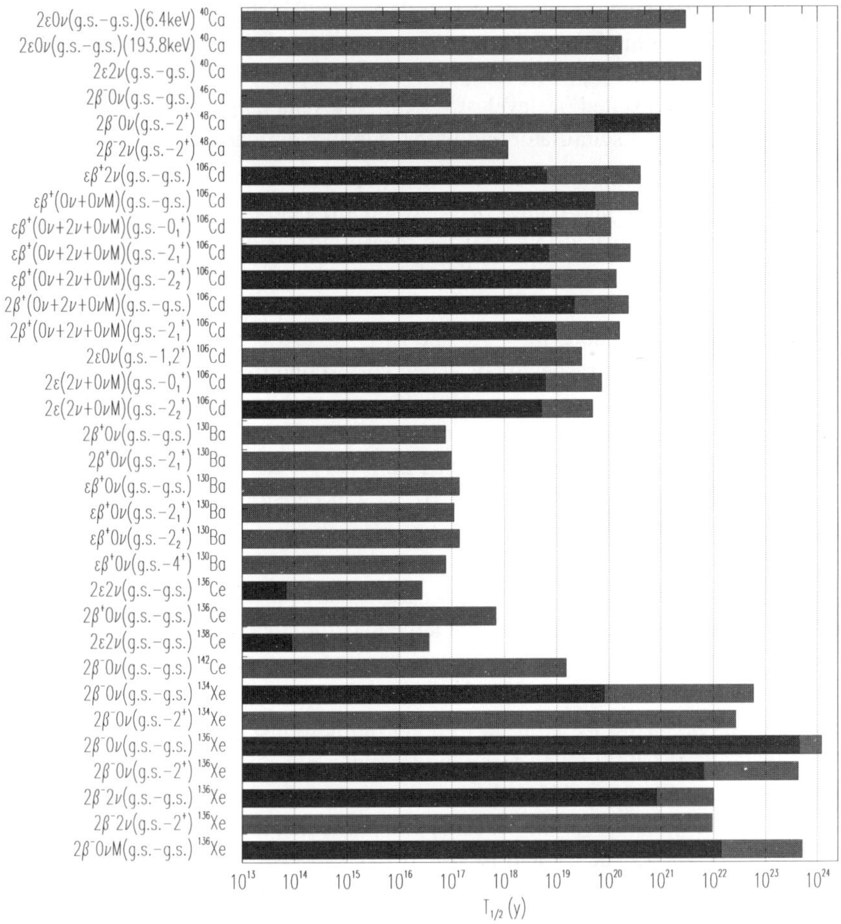

Figure 4. Summary of the $T_{1/2}$ limits obtained by DAMA (dark bars) and by previous experiments (light shaded bars) on various double beta decay processes All limits are at 90% C.L. except the $2\beta^+0\nu$ of ^{136}Ce and $2\beta^+0\nu$ of ^{142}Ce that are at 68% C.L..

or – sometimes – the coincidence technique. Fig. 4 summarizes the many results obtained by DAMA in these searches [37, 38, 42, 43, 44, 45, 46, 47, 48].

In particular, for this purpose the DAMA/LXe set-up [37, 38] and the DAMA/R&D set-up [42, 43, 44, 45, 46, 47, 48] have been used, as well as the low-background Ge in the LNGS facility [50, 51]. Among the obtained results, we remind the search for: i) $\beta\beta$ decay modes in ^{136}Xe and in ^{134}Xe [37, 38]; ii) $\beta\beta$ decay modes in ^{136}Ce and in ^{142}Ce [42]; iii) 2EC2ν decay mode in ^{40}Ca [43]; iv) $\beta\beta$ decay modes in ^{46}Ca and in ^{40}Ca [44]; v) $\beta\beta$ decay modes in ^{106}Cd [45]; vi) $\beta\beta$ and β decay modes in ^{48}Ca [46]; vii) 2EC2ν in ^{136}Ce and in ^{138}Ce and α decay in ^{142}Ce [47]; viii) $2\beta^+0\nu$ and EC$\beta^+0\nu$ decay modes in ^{130}Ba [48]. Moreover, the data taken with a low background $CdWO_4$

scintillator are under analysis and a data taking with a $ZnWO_4$ detector is in progress. Presently, a molybdenum sample of 1 kg mass, enriched in the ^{100}Mo isotope at 99.5%, is in measurement in the low-background set-up with 4 HPGe detectors of the LNGS facility; this experiment [50] will test previous reports on observation of $2\beta 2\nu$ decay of ^{100}Mo to the first excited 0_1^+ level of ^{100}Ru. New data taking are foreseen also e.g. with a new larger BaF_2 scintillator and with Cadmium enriched in ^{106}Cd.

In particular, by using the DAMA/LXe set-up filled with Kr-free Xenon gas containing 17.1% of ^{134}Xe and 68.8% of ^{136}Xe double beta decay modes in ^{134}Xe and ^{136}Xe have been investigated [37, 38]; the measurements have been carried out over 8823.54 h. A joint analysis of the $0\nu\beta\beta$ decay mode in ^{134}Xe and in ^{136}Xe (as suggested in ref. [69]) has also been carried out. The obtained new lower limits for the $0\nu\beta\beta(0^+ \rightarrow 0^+)$ decay mode in ^{134}Xe and in ^{136}Xe (90% C.L.) are $T_{1/2} = 5.8 \cdot 10^{22}$ y and $T_{1/2} = 1.2 \cdot 10^{24}$ y, respectively. They correspond to a limit value on *effective light Majorana neutrino mass* ranging from 1.1 eV to 2.9 eV (90% C.L.), depending on the adopted theoretical model. For the neutrinoless double beta decay with Majoron (M) in the ^{136}Xe isotope the obtained limit is $T_{1/2} = 5.0 \cdot 10^{23}$ y (90% C.L.) and for the $2\nu\beta\beta(0^+ \rightarrow 0^+)$ and the $2\nu\beta\beta(0^+ \rightarrow 2^+)$ decay modes in ^{136}Xe the limits at 90% C.L. are $1.0 \cdot 10^{22}$ y and $9.4 \cdot 10^{21}$ y, respectively. The experimental limit on the $2\nu\beta\beta(0^+ \rightarrow 0^+)$ decay mode is in the range of the theoretical estimate by [70] ($2.11 \cdot 10^{22}$ y) and about a factor 5 higher than that of ref. [71]; on the other hand, similar theoretical estimates suffer of the large uncertainties typically associated to the calculations of the nuclear matrix elements and the overcoming of these theoretical estimates has also been confirmed by the Baksan Xenon experiment [72].

Among the other results we want to point out here just our efforts in preliminary studies on $\beta^+\beta^+/EC\beta^+/ECEC$ decay modes since potentially they can offer both a large number of relatively poorly explored double beta decaying isotopes and the possibility of a good signature for the processes. In particular, neutrinoless double positron ($0\nu 2\beta^+$) decay, β^+/EC decay and EC/EC decay can give, in principle, the same information as $0\nu 2\beta^-$ decay. In addition, they can play a key role in recognising the main production mechanism, strongly depending their calculated half-lives on whether the decay is dominated by Majorana neutrino mass or by right-handed admixtures in weak interactions [5]. On the other hand, even the non-observation of the $0\nu 2\beta^+$ decay would provide very useful additional and complementary information. The current level of experimental sensitivity is modest compared with those already achieved in the $2\beta^-$ decay searches, although the present theoretical expectations for the half-lives of some $2\beta^+$ candidate isotopes are favourable. To reach a suitable sensitivity the use of enriched samples is needed to effectively exploit the potentiality of the $2\beta^+$ processes, which have the advantage of a very clear signature. In fact, e.g., in the $0\nu\beta^+/EC$ decay the monoenergetic positron will give two 511 keV annihilation γ-rays in addition to characteristic X-rays.

A first step in this direction was done in ref. [45] where we present the results of a study on the $2\beta^+$ decay processes in ^{106}Cd, which is a one of the best candidate nuclides because of its rather high Q value ($2\beta^+$), (2771 ± 8) keV and to the favourable theoretical estimates for half-lives. In this experiment, samples of metallic cadmium enriched in ^{106}Cd at $\simeq 68\%$ (total mass $\simeq 154$ g) were placed between two low background NaI(Tl) crystals in the DAMA/R&D set-up; measurements were performed during 4321 h. New improved limits on the halflives of the various decay modes have been obtained [45]; they were in the range (0.3 – 4) 10^{20} y (90% C.L.), that is, they were significantly higher (by factor 6 to 60) than those previously published for the same nuclide, approaching the theoretical estimates. In particular, the calculated $T_{1/2}$ values for the $2\nu\beta^+/EC$ decay of ^{106}Cd are within $(8-40)10^{20}$ y, which are not so far from our limit for this process. These results have strongly supported the interest in pursuing a further enhancement of the experimental sensitivity by additional purification of the enriched cadmium samples and the growth from them of a $^{106}CdWO_4$ scintillator, which could be used as source-detector and, in combination with low background NaI(Tl) crystals, could allow to identify signals. At present some work is in progress in this direction.

Among the other efforts, we remind the results on the neutrinoless double electron capture of ^{40}Ca by using low background CaF_2(Eu) scintillators in the R&D set-up. The obtained new and highly improved $T_{1/2}$ limit: 3.0×10^{21} y, has demonstrated the feasibility and perspectives of developed CaF_2(Eu) scintillators for high sensitive investigations of the double beta decay modes.

On the same line, BaF_2 [48] and CeF_3 [47] scintillators have been used in preliminary investigations both on the level of their residual radiopurity and on the possibility of investigation of some double beta decay modes in ^{130}Ba and of double electron capture processes in ^{136}Ce and in ^{138}Ce.

The potentiality of the coincidence technique to search for double beta decay processes in ^{130}Ba has been studied by using a BaF_2 detector and two low background NaI(Tl) in the DAMA/R&D set-up. The performances of the detector have preliminarily been investigated. Then, as a result, new limits (90% CL) on the lifetimes of the following double beta decay processes in ^{130}Ba have been determined: $\tau^{2\beta^+0\nu(0^+\to0^+)} > 1.1 \times 10^{17}$ y, $\tau^{2\beta^+0\nu(0^+\to2_1^+)} > 1.5 \times 10^{17}$ y, $\tau^{EC\beta^+0\nu(0^+\to0^+)} > 2.0 \times 10^{17}$ y, $\tau^{EC\beta^+0\nu(0^+\to2_1^+)} > 1.6 \times 10^{17}$ y, $\tau^{EC\beta^+0\nu(0^+\to2_2^+)} > 2.0 \times 10^{17}$ y and $\tau^{EC\beta^+0\nu(0^+\to4^+)} > 1.1 \times 10^{17}$ y. The new experimental limits obtained for the $0\nu(0^+ \to 0^+)$ decay modes represent a signicant improvement with respect to those previously available for this isotope, while those involving excited levels have been set for the first time. Measurements with a new BaF_2 detector already at hand are foreseen.

As regards CeF_3, because of the still relatively important residual radioactive contamination, the obtained limits [47] are: $T_{1/2}^{2\nu2K}(^{136}Ce) \geq 2.7(4.5) \times 10^{16}$ y and $T_{1/2}^{2\nu2K}(^{138}Ce) \geq 3.7(6.1) \times 10^{16}$ y respectively, at 90%(68%) C.L. This work has shown that signicant improvements in the sensitivity to rare processes in Ce isotopes can be obtained by exploiting the investigated techniques, but also that it is a crucial point in the achievement of more competitive results to overcome the limitation arising from the residual radioactive impurities in this kind of detector. Purication of scintillators from actinides and their daughters to the level of few μBq/kg could decrease the background at least to few counts/(day keV kg). This together with the use of isotopical enrichment in the production of CeF_3 scintillators would improve the experimental sensitivity at the level expected from the theoretical predictions.

4. Conclusions

Significant results have been obtained by the DAMA collaboration in the investigation of several rare processes and large efforts are in progress to further improve this understanding. New results will be available in near future.

Acknowledgement

I thank the organizer of this workshop for offering me this opportunity of discussion and all the colleagues working in the framework of the DAMA activity.

References

[1] R. Bernabei et al., Phys. Lett. B 389 (1996) 757.

[2] R. Bernabei et al., Phys. Lett. B 424 (1998) 195;

[3] R. Bernabei et al., Il Nuovo Cim. A 112 (1999) 545.

[4] R. Bernabei et al., Phys. Lett. B 450 (1999) 448.

[5] P. Belli et al., Phys. Rev. D 61 (2000) 023512.

[6] R. Bernabei et al., Phys. Lett. B 480 (2000) 23.

[7] R. Bernabei et al., Eur. Phys. J. C 18 (2000) 283.

[8] R. Bernabei et al., Phys. Lett. B 509 (2001) 197.

[9] R. Bernabei et al., Eur. Phys. J. C 23 (2002) 61.

[10] P. Belli et al., Phys. Rev. D 66 (2002) 043503.

[11] R. Bernabei el al., La Rivista del Nuovo Cimento 26 n.1 (2003) 1-73

[12] R. Bernabei et al., Int. J. Mod. Phys. D 13 (2004) 2127.

[13] R. Bernabei et al., Int. J. Mod. Phys. A 21 (2006) 1445.

[14] R. Bernabei et al., Eur. Phys. J. C 47 (2006) 263.

[15] R. Bernabei et al., Phys. Lett. B408 (1997) 439.

[16] P. Belli et al., Phys. Lett. B460 (1999) 236.

[17] R. Bernabei et al., Phys. Rev. Lett. 83 (1999) 4918.

[18] P. Belli et al., Phys. Rev. C60 (1999) 065501.

[19] R. Bernabei et al., Il Nuovo Cimento A112 (1999) 1541.

[20] R. Bernabei et al., Phys. Lett. B 515 (2001) 6.

[21] F. Cappella et al., Eur. Phys. J.-direct C14 (2002) 1.

[22] R. Bernabei et al., Eur. Phys. J. A 23 (2005) 7.

[23] R. Bernabei et al., Eur. Phys. J. A 24 (2005) 51.

[24] P. Belli et al., Il Nuovo Cim. 103A (1990) 767.

[25] P. Belli et al., Astrop. Phys. 5 (1996) 217.

[26] P. Belli et al., Il Nuovo Cim. C 19 (1996) 537.

[27] P. Belli et al., Phys. Lett. B 387 (1996) 222 and Phys. Lett. B 389 (1996) 783 (erratum).

[28] R. Bernabei et al., Phys. Lett. B 436 (1998) 379.

[29] P. Belli et al., Phys. Lett. B 465 (1999) 315.

[30] P. Belli et al., Phys. Rev. D 61 (2000) 117301.

[31] R. Bernabei et al., New J. Phys. 2 (2000) 15.1

[32] R. Bernabei et al., Phys. Lett. B 493 (2000) 12.

[33] R. Bernabei et al., Nucl. Instr. & Meth. A 482 (2002) 728.

[34] R. Bernabei et al., Eur. Phys. J.-direct C 11 (2001) 1.

[35] R. Bernabei et al., in the volume "Beyond the Desert 2003", Springer (2003) 365.

[36] R. Bernabei et al., Eur. Phys. J. A 27 s01 (2006) 35.

[37] R. Bernabei et al., Phys. Lett. B 527 (2002) 182.

[38] R. Bernabei et al., Phys. Lett. B 546 (2002) 23.

[39] R. Bernabei et al., Nucl. Instr. and Meth. A 555 (2005) 270.

[40] R. Bernabei et al., Ukr. J. Phys. 51 (2006) 1037.

[41] P. Belli et al., Nucl. Phys. A (2007) doi: 10.1016/j.nuclphysa.2007.03.001

[42] R. Bernabei et al., Il Nuovo Cimento A110 (1997) 189.

[43] R. Bernabei et al., Astrop. Phys. 7 (1997) 73.

[44] P. Belli et al., Nucl. Phys. B563 (1999) 97.

[45] P. Belli et al., Astrop. Phys. 10 (1999) 115.

[46] R. Bernabei et al., Nucl. Phys. A705 (2002) 29.

[47] P. Belli et al., Nucl. Instr. & Meth. A498 (2003) 352.

[48] R. Cerulli et al., Nucl. Instr. & Meth. A525 (2004) 535.

[49] R. Bernabei et al., in the volume Frontier Objects in Astrophysics and Particle Physics, Vol. 90 (2005) 581.

[50] P. Belli et al., in the volume Nucl. and Atomic Energy, ed. INR-Kiev (2006) 479.

[51] P. Belli et al., Nucl. Instr. and Meth. A 572 (2007) 734.

[52] K.A. Drukier et al., Phys. Rev. D 33 (1986) 3495.

[53] K. Freese et al., Phys. Rev. D 37 (1988) 3388.

[54] A. Bottino et al., Phys. Rev. D 67 (2003) 063519; A. Bottino et al., Phys. Rev. D 68 (2003) 043506.

[55] A. Bottino et al., Phys. Rev. D 69 (2004) 037302.

[56] A. Bottino et al., Phys. Lett. B 402 (1997) 113; Phys. Lett. B 423 (1998) 109; Phys. Rev. D 59 (1999) 095004; Phys. Rev. D 59 (1999) 095003; Astrop. Phys. 10 (1999) 203; Astrop. Phys. 13 (2000) 215; Phys. Rev. D 62 (2000) 056006; Phys. Rev. D 63 (2001) 125003; Nucl. Phys. B 608 (2001) 461.

[57] K. Belotsky, D. Fargion, M. Khlopov and R.V. Konoplich, hep-ph/0411093.

[58] D. Smith and N. Weiner, Phys. Rev. D 64 (2001) 043502; D. Smith and N. Weiner, Phys. Rev. D 72 (2005) 063509.

[59] R. Foot, hep-ph/0308254.

[60] S. Mitra, Phys. Rev. D 71 (2005) 121302.

[61] J.F. Navarro, C.S. Frenk, S.D.M. White, Ap. J. 462 (1996) 563.

[62] J. Diemand, B. Moore, J. Stadel, Nature 433 (2005) 389.

[63] R.A. Ibata, G. Gilmore, M.J. Irwin, Nature 370 (1994) 194.

[64] M. Bellazzini et al., Mon. Not. R. Astronom. Soc. 348 (2004) 12.

[65] F.S. Ling, P. Sikivie, S. Wick, Phys. Rev. D 70 (2004) 123503.

[66] L. Di Lella and K. Zioutas, Astropart. Phys. 19 (2003) 145.

[67] J.D. Lewin and P. F. Smith, Astropart. Phys. 6 (1996) 87.

[68] R. Bernabei et al., to appear on the Proceed. of Vulcano 2006 Workshop, Italy, 2006.

[69] F. Simkovic, P. Domin, A. Faessler, hep-ph/0204278

[70] A. Staudt, K. Muto and H. V. Klapdor, Europhys. Lett. 13 (1990) 31

[71] E. Caurier et al., Nucl. Phys. A654 (1999) 973

[72] Ju. M. Gavriljuk et al., nucl ex/0510071

NEUTRINOLESS DOUBLE BETA DECAY
Editors: V.K.B. Kota and U. Sarkar

SNOLAB: AN OVERVIEW

Rizwan ul Haq

Department of Physics, Laurentian University, Sudbury, ON P3E 2C6, Canada

Abstract

SNOLAB, an international facility for underground science is being built in Sudbury, Ontario, Canada to explore the frontiers of particle and astrophysics (astroparticle physics) with its implicit connections with cosmology and nuclear physics. The facility, started in 2003 and scheduled to be completed in 2007, will be part of a world-wide network of such facilities. Building on the remarkable success of the Sudbury Neutrino Observatory (SNO), SNOLAB experiments will primarily focus on: low-energy solar neutrinos, neutrinoless double beta decay, supernovae detection using neutrinos, and searches for dark matter.

1. Introduction

During the 16th International conference on Neutrino Physics and Astrophysics in 1994, George Marx [1] in his talk entitled – "*The neutrinos grow up*"gave an overview of 20 neutrino conferences and stated at the end – "$\cdots\cdots$*That is why we have to pay attention to daring experimental ideas and new projects. It may well be that for a decade fundamental physics will go underground.*" It is also appropriate to recall what John Bahcall [2] said on the opening of the Sudbury Neutrino Observatory on April 28, 1998 – "*The Sudbury Neutrino Observatory is designed to solve the longest standing scientific mystery: The Case of the Missing Neutrinos.*"

SNO has been remarkably successful [3] in solving this long standing Solar Neutrino problem by showing that – (i) the measurement of the total ^{8}B neutrino flux from the Sun is in excellent agreement with the standard solar model and (ii) approximately 2/3 of the electron neutrinos from the Sun transform (neutrino oscillations) to other neutrino types on their way to Earth. With the development of experimental techniques and theoretical advancements, there is a great deal of interest and excitement in astroparticle physics worldwide as we probe some of the fundamental questions in physics and cosmology regarding the nature of the universe.

With the research experience of SNO and the investments made therein, it is natural that the international underground science facility, SNOLAB, be built at the same location as the deepest and cleanest underground laboratory in the world at a relatively modest cost. Situated at the 2020 m level of the CVRD-INCO Creighton Mine in Sudbury, it provides excellent shielding from cosmic rays and a low-background (ultra-low radioactivity) environment particularly conducive to highly sensitive experiments. Initial funding has been provided by the Canada Foundation for Innovation (CFI), Ontario Innovation Trust (OIT), the Northern Ontario Heritage Fund Corporation (NOHFC), FeDNoR and the City of Greater Sudbury.

The aim of SNOLAB is to establish a world-class facility for the next generation of experiments that will address fundamental questions in particle physics, astrophysics and cosmology. Experiments at SNOLAB will deal with the following areas: (i) low-energy solar neutrinos, (ii) neutrinoless double beta decay, (iii) supernova detection using neutrinos and (iv) search for dark matter.

The scientific community at large has been engaged in defining the physics program as well as the design of SNOLAB through a series of workshops. The 1st SNOLAB Science Workshop was held in Sudbury in August 2002 with the aim of introducing the facility and developing the physics program. Scientific discussions dealt largely with the kind of experiments that could be best conducted at SNOLAB as well as the design and planning of this new facility. The 2nd Science Workshop held in Ottawa in November 2002 discussed prospective experiments and their infrastructure needs. Topics covered were: Solar Neutrinos, Dark Matter, Double-Beta Decay, Low Level Counting and Screening as well as the management structure and the scientific review process to be put in place at SNOLAB.

The 3rd Workshop on Underground Science was held in Sudbury in May 2004 with the aim of refining the infrastructure needs of the facility and to start the experiment evaluation process which would result in the best experimental program for SNOLAB. The 4th Workshop on Underground Science and Experiments for SNOLAB took place in Sudbury in August 2005. The goal was to define the scientific program through recommendations regarding an initial set of experiments. The 5th Workshop on Science and Experiments for SNOLAB held in Sudbury in August 2006 evaluated the status and progress on the development of the facility as well as the progress and schedule of prospective experiments. It also considered the long term scientific roadmap of SNOLAB.

SNOLAB Workshop talks may be found at www.snolab.ca/workshops/

2. The Underground Facility

SNOLAB facilities will provide 30,000 sq ft of clean room space underground and 34,000 sq ft of laboratory, office and meeting space on the surface. The underground laboratory will consist of the existing (72 ft diameter, 90 ft high) SNO Cavern, a 50'× 60' × 50' rectangular hall, a Cryo-pit of dimensions 50'× 50' and Ladder Laboratory Area (180' × 20'× 12' + 300' × 15'× 10'). The underground space will be separated into rooms for each experiment with the possibility of substantial isolation between experiments. It is expected that there will be space for ten 100-tonne experiments, several 1-tonne experiments as well as dedicated areas for prototyping. The ladder area will provide space for a Chemistry Lab, a low-background counting lab, an electronics lab and a clean machine shop [4].

The surface building, completed in November 2005, has been designed to provide laboratory and infrastructure support to the experiments. It consists of various cleanroom laboratories for development, assembly and testing, computing facilities, conference and meeting rooms, and office space for on-site visiting staff and scientists [4]. Further, facilities for material testing and prototyping to support the underground experiments are being set up in the department of physics at Laurentian University [5]. These facilities will include an ongoing radiotracer testing program since, in accordance with the policy to prevent contamination, no radiotracer studies involving liquid radioisotope sources may be done at the SNOLAB surface or underground labs.

3. SNOLAB Organization

The facility is managed by the Sudbury Neutrino Observatory Institute (SNOI) and a SNOI Board, which includes representatives of the participating universities and research laboratories, appoints the Director [6]. The scientific organization of SNOLAB consist of several structures namely SNOLAB operations management, the Scientific Executive Committee (SEC), the Experiment Advisory Committee (EAC), the Technical Review Committee (TRC) and a SNOLAB Users Group (SUG). The aim is to provide a mechanism for: evaluating experimental proposals based on scientific merit and feasibility; interaction for the SNOLAB experiments with management and with each other; creating an environment for the user community to interact, and perhaps most importantly, ensuring a safe working environment [7].

3.1. Scientific Executive Committee (SEC)

The SEC is composed of SNOLAB senior management, representatives of the SNOLAB institutions and the principal investigators from the experiments. The principal objective of the committee is to provide interface with the international scientific community and to assist the SNOLAB director in meeting the needs of on-going experiments [7]. It also provides guidance to the director regarding the scientific program and helps to resolve conflicts between experiments.

3.2. Experiment Advisory Committee (EAC)

The EAC will review the Letter-of-Intent (LOIs) received from the scientific community and evaluate the merit and technical feasibility of the experiments proposed for SNOLAB. It will serve as an expert panel and provide advice to the director in setting priorities for experiments to be carried out in the laboratory. The goal of EAC is to develop the most promising program taking into account the scientific potential of the project; the experiment must have significant discovery potential, take advantage of the great depth and other attractive attributes of SNOLAB, and should be determined in the context of other plans worldwide [8]. It is composed of experts from the particle astrophysics community [7].

3.3. Technical Review Committee (TRC)

The TRC will assess the technical readiness of the project for deployment as well as the infrastructure needs of the experiment. It will also ensure regulatory approval. The members of this committee are technical experts from SNOLAB. External experts may be used when necessary [7].

3.4. SNOLAB Users Group (SUG)

The SUG has been formed to provide a forum for discussion and information dissemination regarding the development of SNOLAB and its scientific program. It is composed of scientists, engineers and students involved with the facility [7].

4. The Physics Program

The scientific program for SNOLAB, as mentioned earlier, is being developed through a series of workshops. The program is designed to exploit the unique features of the facility. SNOLAB with its large depth and extreme cleanliness will provide the ideal laboratory for experiments dealing with some of the fundamental question in physics and particle astrophysics such as: (i) what are the masses of the three neutrino species and how do they transform from one to the other? (ii) are there other types of neutrinos? (iii) how can dark matter be detected and what is its composition; how much of it is neutrinos? These questions while focused on a better understanding of neutrinos are expected to have a much larger bearing on the scientific landscape [7].

4.1. Solar Neutrinos

It is clear that SNO has resolved the so called "*Solar Neutrino Problem.* " Precision measurements of the total flux of ^{8}B solar neutrinos agree well with the prediction of the Solar Model. Further, it has been shown through the neutral current (NC) and charge current (CC) measurements that about 2/3 of the electron neutrinos change to other types in transit to Earth. The latest SNO results as well as other measurements support the hypothesis of finite neutrino mass as the dominant mechanism for oscillations and flavour change and these results, as well as those of the SuperKamiokande [8] experiment, allow mass differences between neutrino species to be determined. These measurements of other flavours and confirmation of neutrino oscillations has opened the door for exciting future solar neutrino experiments. It is clear that beyond these significant results lies a lot of interesting physics related to neutrinos and their interaction with matter. Of particular interest are the determination of the absolute mass scale, the origin of flavour transformations and the mixing pattern.

A capability enhanced new solar neutrino experiment using SNO infrastructure could help to determine properties of solar neutrinos leading to a better understanding of energy production the Sun. Plans are being implemented to use the SNO detector for precision measurements of low-energy neutrinos from the Sun by replacing the heavy water with an organic liquid scintillator (SNO+) thereby lowering the neutrino energy threshold and observing neutrinos from different solar nuclear reactions. SNOLAB is perhaps the only existing deep site where pep solar neutrinos could be measured with precision. The depth eliminates problems associated with cosmic-ray induced radioactivity such as ^{11}C. There is interest in the study of pep solar neutrinos since such measurements will provide information about the neutrino oscillation parameters, in particular, regarding the mixing angle [9].

Looked at as a multipurpose detector, SNO+ could also be used in the measurement of electron neutrinos arising due to the natural radioactivity of Uranium and Thorium in the Earth. These measurements of geo-neutrinos would determine the amount of Uranium and Thorium in the Earths crust and thus provide important data for geological models, in particular for those dealing with the chemical origin and thermal history [9].

4.2. Double Beta Decay

The search for neutrinoless double-beta decay ($0\nu\beta\beta$), the topic of this meeting, is important for several reasons: (i) to establish whether neutrinos are Majorana or Dirac particles; (ii) to test Lepton number conservation; (iii) to establish the absolute neutrino mass scale and along with the mass difference information from oscillation experiments, provide information on neutrino masses and their mass hierarchy.

Given the fact that these experiments will have low event rates, SNOLAB with its depth, cleanliness and low background provides an excellent site for conducting them. Measurement of the decay rate by these experiments can provide the best information on neutrino mass. However, it should be pointed out that the half-life also depends on the Fermi and Gamow-Teller nuclear matrix elements. These matrix elements are difficult to calculate due to the size of the shell model space (see for example [27]) and uncertainties in their approximate calculations will show up in the neutrino mass uncertainties. It is pleasing to note that theoretical activity in the calculation of double beta-decay rates is on the rise, in part due to the on-going as well as future experiments with increased sensitivity.

Several excellent experiments are either on-going or have been proposed. There is the possibility of exploring double beta-decay in SNO by placing in the SNO acrylic vessel a variety of nuclei [10]. It has been suggested [11] that the SNO detector filled with scintillator offers a unique opportunity to study $0\nu\beta\beta$ of Xenon. Thus SNO with liquid scintillator plus double beta isotopes, SNO++, could be used as a detector. The most promising isotope is ^{150}Nd (1% natural Nd-loaded liquid scintillator in SNO++ has 560 kg of ^{150}Nd) [12]. Although the energy resolution of the liquid scintillator is poor, it is expected that high statistics would make up for this drawback and permit a sensitive test of the neutrinoless decay mode.

The Enriched Xenon Observatory (EXO) project has proposed a larger detector at SNOLAB. It aims to search for $0\nu\beta\beta$ using Xenon [13]. The advantages of using Xe are: (i) ease of purifying a noble gas; (ii) can be re-purified during experiment; (iii) no need to grow crystals; (iv) absence of long-lived radio-isotopes; and (v) the daughter Barium could be tagged on an event by event basis. Both a liquid as well as a possible gas version of the experiment are being explored.

The Majorana project proposes to build a large Germanium (^{76}Ge) detector in a copper cryostat [14]. The experiment has several advantages namely: large mass, natural isotope, good energy resolution, ease of operation, demonstrated technology, and, above all, the source also serves as the detector. Another ^{76}Ge detector experiment underway at Underground National Laboratory of Gran Sasso (Laboratori Nazionali del Gran Sasso (LNGS)), Italy is the Germanium Detector Assembly (GERDA). This $0\nu\beta\beta$ experiment with a cryogenic liquid shield is planned to (i) verify the concept of a liquid shield, (ii) check the existing claim [28], and (iii) improve germanium detectors [15].

Yet another experiment for the determination of the effective Majorana neutrino mass through double beta decay is COBRA which has been running at LNGS since August 2003. The experiment searches for $0\nu\beta\beta$ in Te, using semiconductor CdZnTe detectors with the following advantages: source is the detector, energy resolution is good, the experiment is done at room temperature, industrial development of CdTe detectors exists and it has a modular design [16].

4.3. Supernovae

There is interest in the detection of neutrinos from a supernova. The study of neutrinos produced in the gravitational collapse and subsequent explosion of a massive star can reveal much about the dynamics of the process.

Supernova neutrino burst monitoring was carried out in the SNO experiment and SNO was a founding member of the Supernova Early Warning System (SNEWS) which included underground neutrino laboratories around the world. It is expected that this detection capability will continue with SNOLAB; the liquid scintillator in the SNO+ acrylic vessel [9] would serve as the supernova neutrino monitor.

Another possibility for a supernova detector being explored is the Helium and Lead Observatory (HALO) [17, 18]. The basic idea is to exploit the high neutrino cross section of Lead (^{208}Pb) and detect the emitted neutrons using Helium (^{3}He) proportional counters.

$$\nu_e + {}^{208}Pb \rightarrow {}^{207}Bi + n + e^-$$

$$\nu_e + {}^{208}Pb \rightarrow {}^{206}Bi + 2n + e^-$$

$$\nu_x + {}^{208}Pb \rightarrow {}^{207}Pb + n$$

$$\nu_x + {}^{208}Pb \rightarrow {}^{206}Pb + 2n$$

HALO'ss sensitivity to electron neutrinos and neutral current neutrino processes will be valuable since other experiments are primarily sensitive to electron antineutrinos.

The advantages of this low-cost, low maintenance and long-term detector are that most of the material is already available in Sudbury, the hardware has been tested and the facility can be easily expanded.

4.4. Dark Matter

It is generally believed that a significant fraction of the matter in the Universe is dark. Currently, there is a great deal of interest in trying to understand dark matter and novel methods are being proposed to search for dark matter particles. Neutrinos from the Big Bang and astrophysical sources contribute to hot dark matter, but upper limits on their masses indicate that other dark matter particles are likely. The basic premise in most experiments searching for the Weakly-Interacting Massive Particle (WIMP) is to study their interaction with matter and measure the subsequent nuclear recoil. The experimental challenge is to be able to discriminate this event from the background since event rates are low and recoil energy rather small.

The Project in Canada to Search for SuperSymmetric Objects (PICASSO) searches for dark matter through the direct detection of WIMPs via their spin-dependent interaction with nuclei [19, 20]. The detector medium is Freon (C_4F_{10}) and the experiment is designed to observe the recoil of ^{19}F nuclei which has the most favourable cross section. The PICASSO detectors are based on the bubble chamber principle. In this technique the detecting medium is tiny droplets of a superheated liquid embedded in a polymer matrix. The liquid is kept in a metastable state until a heat spike, produced by energy deposited from particle interactions (e.g. nuclear recoils), triggers a phase transition [29]. The droplets transform to bubbles of vapour that remain trapped in the surrounding polymer. The phase transition is accompanied by an acoustic signal which is recorded by piezoelectric transducers. These detectors are reusable since the bubbles can be recompressed into droplets by increasing the pressure on the polymer matrix.

The PICASSO Collaboration has been developing the superheated droplet technique for the search of dark matter [29]. The main advantages of this technique are: (i) operating temperatures are close to room temperature and (ii) in the operating temperature range the detectors are sensitive neither to gamma and beta radiation nor to cosmic ray muons. Further, the collaboration fabricates its own detectors.

In 2005, successful completion of the proof of principle tests with three 1-liter detectors ($\sim$ 21g of active mass) led to one of the best limits for the spin dependent sector [30]. The Collaboration is now installing the next phase of the experiment at SNOLAB. The experiment will have thirty-two 4.5 liter detectors (2.6 kg of active mass). This phase is part of a 3-phase timeline for the future development of the project in which the active mass will eventually increase to 100 kg. This should result in reduced background and increased sensitivity of the experiment.

The Dark Matter Experiment with Argon PSD (DEAP) is designed to detect spin-independent WIMP-nucleon scattering in liquid Argon (^{40}Ar) [21]. The spherical liquid Argon detector is surrounded by photo multiplier tubes to detect scintillation photons. The basic principle is "—*the possibility of using the differences in scintillation photon times for electrons and nuclear recoils in liquid argon for discrimination of backgrounds in a WIMP search*"[22]. The experiment can be scaled up and whereas DEAP-0 (1 kg detector) at Los Alamos National Laboratory has been successfully completed, plans and assembly are underway for DEAP-1 (10 kg) to DEAP-3 (1000 kg) to be situated at SNOLAB. Advantages of using liquid argon are that it is inexpensive, provides good light yield, good pulse shape discrimination, and can be easily purified.

Other experiments dealing with the search for dark matter are: the Cryogenic Dark Matter Search (CDMS) [23], the ZEPLIN Dark Matter program [24], Xenon dark matter experiment [25], and DRIFT [26].

4.5. Special considerations for experiments underground

There are certain design criteria for experiments being conducted underground. With the experience at SNO, one may regard these as a set of design rules. Particular attention should be paid to the following – (i) redundant communications with a view to minimize single point failures, (ii) remote control of the experiment such as start up, shut down, control and monitoring should be maximized, (iii) systems underground must be designed such that they can run independently and without intervention from surface, (iv) in the event of communications failure, the system should go to a safe state, and (v) experiment design should take into account the seismic environment [4]. The 25% higher pressures and presence of radon in the laboratory air must also be allowed for.

5. Conclusions

SNOLAB has been created to serve as a unique laboratory for research leading to "new"physics. The aim is to develop a scientific experimental program of the highest merit and the experiments at SNOLAB will go a long way in answering some of the fundamental questions in astroparticle physics. SNOLAB will be part of a worldwide network of facilities for underground science, facilities that exist in Italy, Japan and Russia. To paraphrase George Marx – "*it appears likely that for the next several decades, key progress in fundamental physics will remain underground.*"

Acknowledgements

I am grateful to Dr. Jacques Farine and Dr. Doug Hallman for reading the manuscript and for their valuable suggestions. I would also like to thank Dr. Ubi Wichoski for discussions regarding the PICASSO project.

References

[1] George Marx, Nucl. Phys. B (Proc. Suppl.) **38**, 518 (1995).

[2] John Bahcall, SNO and the Missing Solar Neutrinos, talk delivered on the occasion of the opening of the Sudbury Neutrino Observatory, April 28, 1998.

[3] The SNO Collaboration, Phys. Rev. Lett. **87**, 071301 (2001); Phys. Rev. Lett. **89**, 011302 (2002); Phys. Rev. Lett. **89**, 011301 (2002); Phys. Rev. Lett. **92**, 102004 (2004); Phys. Rev. Lett. **92**, 181301 (2004); Phys. Rev. **C 75**, 045502 (2007). Participating Institutions in SNO: Queens University, Carleton University, University of Guelph, Laurentian University, University of British Columbia, University of Pennsylvania, Los Alamos National Laboratory, University of Washington, Oxford University, Brookhaven National Laboratory, TRIUMF, University of Texas, Indiana University, South Bend and Massachusetts Institute of Technology. Earlier participation by Chalk River Laboratories, National Research Council of Canada, University of California at Irvine and Princeton University.

[4] Fraser Duncan, Lab Design and Infrastructure, SNOLAB Workshop III (2004); Construction and Development Status, SNOLAB Workshop V (2006).

[5] Jacques Farine, Facilities at Laurentian University, SNOLAB Workshop II (2002); SNOLAB Workshop III (2004).

[6] The SNO Institute University partners are Carleton University, University of Guelph, Laurentian University, Queens University, University of British Columbia and Universit de Montral. The SNOLAB Director is Dr. David Sinclair, Carleton University.

[7] Fraser Duncan, SNOLAB Operations and Organization, SNOLAB Workshop V (2006); Andrew Hime, Feedback from the SNOLab Scientific Executive Committee, SNOLAB Workshop III (2004); Barry Barish, SNOLAB Experimental Advisory Committee, SNOLAB Workshop III (2004).

[8] S. Fukuda et. al., Phys. Rev. Lett. **86**, 5651 (2001); Phys. Rev. Lett. **86**, 5656 (2001).

[9] M. Chen, SNO with Liquid Scintillator, SNOLAB Workshop III (2004); Mark Chen, SNO+: SNO with Liquid Scintillator, SNOLAB Workshop IV (2005).

[10] Art McDonald, Double Beta Decay in SNO, SNOLAB Workshop III (2004).

[11] Barbara Caccianiga, Neutrinoless double beta decay with Xenon, SNOLAB Workshop III (2004).

[12] Mark Chen, SNO+, SNOLAB Workshop V (2006).

[13] David Sinclair, EXO at SNOLAB, SNOLAB Workshop III (2004); SNOLAB and EXO, SNOLAB Workshop IV (2005); A Gas Option for EXO, SNOLAB Workshop V (2006).

[14] Steve Elliott, The Majorana Project, SNOLAB Workshop III (2004); Peter Doe, Majorana double beta decay program, SNOLAB Workshop IV (2005); A. L. Hallin, The Majorana Neutrinoless double -decay experiment, SNOLAB Workshop V (2006).

[15] I. Abt, GERDA: GERmanium Detector Assembly, SNOLAB Workshop III (2004); Michael Altman, Investigating Neutrinoless Double Beta Decay; Status of the GERDA Experiment, SNOLAB Workshop IV (2005); Allen Caldwell, GERDA: The GERmanium Detector Array for the search for neutrinoless $\beta\beta$ decays of ^{76}Ge, SNOLAB Workshop V (2006).

[16] K. Zuber, COBRA: A new approach to double beta decay, SNOLAB Workshop II (2002); COBRA, SNOLAB Workshop III (2004); COBRA, SNOLAB Workshop IV (2005); COBRA, SNOLAB Workshop V (2006).

[17] Charles Duba, HALO: Helium and Lead Observatory, SNOLAB Workshop III (2004).

[18] C. J. Virtue, HALO – a Helium and Lead Observatory, SNOLAB Workshop IV (2005); SNOLAB Workshop V (2006). Collaborating Institutions in HALO are: University

of Washington, Laurentian University, Los Alamos National Laboratory, Lawrence Berkeley National Laboratory, Carleton University, Queens University, Duke University and University of Minnesota Duluth.

[19] V. Zacek, The Picasso Collaboration, SNOLAB Workshop III (2004); Picasso – Searching for Dark Matter, SNOLAB Workshop V (2006). Collaborating Institutions in Picasso are: Universit de Montral, Queens University, Laurentian University, University of Indiana, South Bend, Czech Technical University of Prague, Bubble Technology Industries, and an MOU with SIMPLE collaboration (Universit de Paris VI & VII and Universita de Lisboa).

[20] Tony Noble, The Picasso Dark Matter Search Experiment, SNOLAB Workshop IV (2005).

[21] Mark Boulay, DEAP: Dark Matter Experiment with Argon PSD, SNOLAB Workshop IV (2005), Progress on DEAP, SNOLAB Workshop V (2006). DEAP-1 Collaboration: Queens University, Carleton University, Case Western Reserve University, Los Alamos National Laboratory, SNOLAB and Yale University.

[22] M. G. Boulay and A. Hime, Direct WIMP Detection Using Scintillation Time Discrimination in Liquid Argon, http://arxiv.org/astro-ph/0411358

[23] Dan Akerib, The Cryogenic Dark Matter Search: Status and Future Plans, SNOLAB Workshop III (2004); SuperCDMS, SNOLAB Workshop IV (2005); Dan Bauer, SuperCDMS, SNOLAB Workshop V (2006).

[24] Nigel Smith, The ZEPLIN Dark Matter Programme, SNOLAB Workshop III (2004); Hanguo Wang, ZEPLIN: A Dark Matter Direct Search Program Using Liquid Xenon, SNOLAB Workshop IV (2005); Tim Sumner and Hanguo Wang, ZEPLIN, SNOLAB Workshop V (2006).

[25] T. Shutt, XENON dark matter experiment, SNOLAB Workshop III (2004); XENON, SNOLAB Workshop IV (2005) and Workshop V (2006).

[26] C. J. Martoff, DRIFT – Past, Present, Future, SNOLAB Workshop III (2004); Sean Paling, DRIFT III, SNOLAB Workshop IV (2005).

[27] E. Caurier, F.Nowacki, A. Poves, J. Retamosa, Nucl. Phys. **A 654**, 973 (1999).

[28] H. V. Klapdor-Kleingrothaus, I. V. Krivosheina, A. Dietz, O. Chkvorets, Phys. Lett. **B 586**, 198 (2004).

[29] M. Barnabe-Heider et. al., Nucl Inst. and Meth. **A 555**, 184 (2005)

[30] M. Barnabe-Heider, et al., Phys. Lett. **B 624**, 186 (2005).